A WHITE PINE EMPIRE

THE LIFE OF A LUMBERMAN

By

JOHN EMMETT NELLIGAN

NORTH STAR PRESS
St. Cloud, Minnesota

Dedication

To My Daughter Helen Nelligan Geartts from Father John Emmett Nelligan

January 1st 1930

Publisher's Note to the Second Edition

It is difficult for an editor to criticize the writings of an earlier time. The style and syntax of fourty years ago are far from acceptable in our day. All too many of the latter day phrases sound over-blown and triumphant to our very casual modern ear. However, to meddle with the writings of an early ghost would deprive the reader of some of the flavor of the times, and the pace of the readers of the twenties.

As a result of our decision to not re-write the words of "Ghost" Sheridan, the character of John Nelligan needs this prefatory note to explain his character a bit.

John Sheridan wrote from the notes and conversations of John Nelligan. Sheridan suffered from a false sense of social responsibility, and felt that he must make John Nelligan more of a gentleman in print. He overlooked the fact that John Nelligan was one of nature's own gentlemen. John Nelligan was a natural man, good, hard-working, fun-loving and successful at his chosen vocation, lumbering. John Nelligan was an earthy and direct man, one used to action, and comfortable in the role of command and leadership.

Thus, the modern reader of this re-published version of Nelligan's autobiography must read the story with an eye cocked between lines to get the real flavor of Nelligan and his times in the pine woods of Michigan, Wisconsin and Minnesota.

John Nelligan, was in fact, a direct and somewhat profane man who was given to enjoying the life he was gifted to live. He enjoyed his work greatly, and thrived in the give and take of working in a rough and tumble industry in its infant days here in the United States. Nelligan loved the outdoor life working with his crews of hard-muscled jacks. He was a driver who wanted and got lumber production from his men, but who enjoyed the pranks and jokes of his times. We might look upon some of their antics as crude, rough and dangerous. I think that we must remember that life in the woods was hard, and dull but dangerous and when the men had a chance for a joke, or some playful relaxation they went after it "hammer and tongs".

In conversations with Nelligan's daughter, son-in-law, and old friends, we got to know the character of the man better. We began to know more fully a man who even in his middle eighty years was full of interest in the events of the day. Visits to the downtown area of Escanaba, Michigan, and Milwaukee, Wisconsin, to talk with his friends would keep him abreast of the changes in the lumbering business of the region. In total, John Nelligan was a man of his times. He was the kind of man who contributed his enthusiasm, natural talents, and ambition to make a success of living in America. His is the story of thousands of immigrants in the later years of the 19th Century, who brought their ability and desire to succeed and who really made America live from shore through woods and plains to distant shore.

It is our hope that by bringing this book back to print after forty years the readers in the latter half of the 20th century will learn to really know and appreciate those who plowed, sawed and branded our land into a readiness for our comfortable modern technical society where we, unfortunately, have too few John Nelligans.

TABLE OF CONTENTS

THE LIFE OF A LUMBERMAN

By

JOHN EMMETT NELLIGAN

as told to

CHARLES M. SHERIDAN

FOREWORD

THE old pine forests of Michigan and Wisconsin have
practically vanished into the mists of the past. The
unique and heroic race of men who worked among them have
also almost gone out of existence. Undoubtedly their prog-
eny carry the energy and the pioneer spirit of their fathers
into other lands and other fields of activity, but only here
and there do the genuine old timers linger. The tale of these
men, their deeds and doings, has almost become legendary.
Like most legends, that of the cowboy occurs to me at the
moment, the true characteristics are likely to be overlaid by a
sort of second hand romance or an equally second hand real-
ism spread over them by tellers and writers who know nothing of themselves concerning these conditions. Therefore,
it is always a great pleasure for one to see American actualities preserved, to know that one who has been an actual part
of the time is telling about them.

I knew Mr. Nelligan in the old days of Michigan's and
Wisconsin's supremacy in the lumber business, and I re-

member him as one who was representative of all the qualities that made the times great. It is therefore with the greatest pleasure that I learn he is to place his reminiscences on paper. He knew and was a great part of that old life, and his rendering of it should be not only valuable to the student of history, but also should give the greatest pleasure as affording a bona fide glimpse of real conditions that have gone.

Little Hill, STEWART EDWARD WHITE
Burlingame, California,
February 22, 1927.

INTRODUCTION

JOHN E. NELLIGAN'S STORY

ONE day in the first week of April (1929) a packet of thirty-four long envelops came to my office. Each of these contained a few typewritten sheets. The manuscript was carbon copy on glaring orange colored paper, which having been folded twice for enclosing, from purely physical causes was difficult to read, notwithstanding its pleasing literary quality. A letter from a young Milwaukee business man advised that he had sent the material, which constituted a book on lumbering. This, he said, had been written by a neighbor, Mr. John E. Nelligan, who was seventy-seven years of age and had spent half a century in the pine forests as a practical lumbering and logging operator. He felt that the story was significant in relation to the history of an industry which was passing, and that Mr. Nelligan, in view of his meagre opportunities for securing an education, had succeeded well in the telling of it.

The first rapid reading of the chapters convinced me that most of the matter was genuine, but that there was a "ghost writer" in the background who, rather than the Irish lumberman, was responsible for the literary character of the work and also for some minor portions of the narrative which appeared to have been added for the sake of completeness. My suggestion that Nelligan must have had a collaborator gave my correspondent a little shock, for he appears to have been guilelessly unsuspicious on that head. But when he next saw Nelligan, he was frankly told that a young writer of Washburn, Wisconsin, had put the material in form.

The interview between the aged lumberman and his youthful neighbor seems to have gone awry in some way: for the latter complained to me that he ought to have been informed earlier about the literary author, washed his hands of the whole affair, and suggested that I write Nelligan directly. This I did, telling him that I had found the manuscript interesting, but that I detected in it a strain of artificiality which seemed to indicate that someone had been permitted to work over his matter freely, and requested him to inform me whether or not he had in his possession the original notes from which the book had been written. His reply came promptly, under date of April 22, 1929. In it Mr. Nelligan says, "Charles M. Sheridan reconstructed the manuscript—*The Life of a Lumberman*—for me." He promises to call on me at my office very soon to talk over the whole question of publication. He thinks the chapter on reforestation (which I was sure had been written by someone else) will be valuable for the State. The letter closes in this somewhat unexpected if not startling vein: "Mr. —— (naming the Milwaukee man referred to above) tried to take advantage of me in wanting me to prepare an agreement for him, also to give him my power of attorney. I told him to go to hell, and I think he is on the way there, as he left Milwaukee early this morning.

Very truly yours,

JOHN EMMETT NELLIGAN."

Next day I was informed by long distance phone that Mr. Nelligan would call at my office on the morrow. He came to Madison as agreed. Naturally, I was much interested in meeting so virile an oldster, who is described in the narrative as a husky, fighting Irish woodsman, six feet three in height. The first glimpse of his still vigorous, erect physique, convinced me that the personal description contained in the manuscript was correct, though it would not have helped me to "put backs with him," in the way Lincoln was always doing when tall men called upon him, for my own modest five feet eight and a half could not possibly have served as a measure. His first remark, after the meeting, showed that he had all of the Irishman's responsiveness to environment. Quoth he: "I didn't know but you might be offended at what I told you I said to that young man." Having assured him that a picturesque form of speech was not wholly strange to me, we got down to the business in hand.

He brought me what he described as the original draft of the book, by which he meant the fair copy. This was a bound volume of manuscript, neatly typed on white bond paper. It contained, aside from the author's photo and half a dozen lumbering scenes, a foreword by Stewart Edward White, a note by the collaborator, Charles M. Sheridan, a table of contents, and 195 pages of closely typed text distributed among thirty-four chapters.

In addition, Mr. Nelligan left with me all the letters received by him from his collaborator. These are delightfully revealing. They show that Nelligan had harbored for some time the ambition to write his experiences. Knowing that he would require literary aid, and having read a story in *Columbia* magazine called "Kings of the White Water," he decided its author would be the right man to do the work. So he wrote to him. Mr. Sheridan replied favorably. "Although," he says, "I have had little direct contact with lumbering activities, aside from the sport of logrolling, I have read and heard much about the industry, and, living as I do in a once thriving lumber community among many old timbermen, I am passably well versed in the lore of the game." In November, 1926, Nelligan visited Sheridan at Washburn, leaving with him a quantity of rough notes. He had absolutely no records save those graven upon his memory which, however, was extraordinarily retentive. Nelligan had no idea of literary form or of proportion, and it would be Sheridan's duty to supply these deficiencies.

The latter's first step, after contracting to do the work, was to study the raw notes, block out from them a tentative outline of chapters, and then secure from the author the necessary material to round the stories out. For the latter purpose he spent several days with Nelligan in Milwaukee, industriously questioning him and recording his detailed reminiscences. Then he began writing and as chapters were whipped into form he sent drafts to the author for further suggestions and for the correction of errors. In this manner the writing proceeded, Sheridan devoting most of his time to it for more than a year, and being at some pains to convince his principal that he must not hurry the job unduly. At last on the seventh of January, 1928, the final revision in fair copy—the one which was left with me—was mailed to Mr. Nelligan.

The correspondence proves that the work was carried out conscientiously by both author and collaborator. Some parts, like reforestation, were wholly the work of the latter, who obtained his material from standard treatises on that subject. Sheridan, too, spurred Nelligan to try to remember matters which would prove interesting to the reader. Paul Bunyan tales are a case in point. Apparently Nelligan, despite his diversified experience in the woods of New Brunswick, Maine, Pennsylvania, Michigan, and Wisconsin, had hardly so much as heard of the redoubtable Paul or of his blue ox, Babe. Why this was so we can only surmise. Perhaps, it was because he employed so largely foreign born —especially Irish—lumberjacks, instead of natives. At all events he could not remember anything worth while and at the good last Sheridan was forced to take a few items out of a book; for Bunyan stories belonged in the chapter on camp recreation!

With the aid of the letters it proved an easy task to identify the materials which were extraneous to Nelligan's personal experience. These—aside from the few harmless Bunyan yarns—I excluded. Some other matter which, though genuine, was not significant, or distinctive of the lumbering business, was likewise cut out. In these ways the manuscript was reduced to about ninety per cent of its original bulk. I also modified the order slightly, assembled it under fewer chapter heads, and organized the chapters under three nearly equal *Parts* for production in successive numbers of the *Wisconsin Magazine of History.* The collaborator's note was omitted at his request.

Mr. Nelligan's story, charmingly written as it has been by his capable literary sponsor, is an exceptionally comprehensive addition to our knowledge of the processes of logging and lumbering; because, unlike such a writer as George H. Warren, author of *The Pioneer Woodsman as He is Related to Lumbering in the Northwest,* Nelligan was land looker, woods-foreman, river boss, contractor, and, in fact, everything except a sawmill operator. The book constitutes, also, a genial commentary by a shrewd observer, upon the woods life of the Great Lakes states during their greatest lumbering period. For these reasons we are glad to present it to the readers of the magazine—without expurgating its spiritually disinfected profanity which is an unconscious revelation of one phase of that life.

It has been a pleasure to help open a way for the production of the story in book form. Like Stewart Edward White, we are glad to see "American actualities preserved, to know that one who has been an actual part of the time is telling about them."

JOSEPH SCHAFER

PART I

GROWING UP TO BE A LUMBERJACK

Chapter I   Beginning in New Brunswick

My father's name was Patrick Nelligan, and if I remember correctly what I was told as a boy, he and his kin were natives of County Limerick in Ireland. The years of his youth found him in the little port of Dingle, which lies on Dingle Bay on the west coast of Ireland in County Kerry, and there he met the girl who was to become his bride and my mother, Johanna Sullivan. At the time of their marriage many eyes in Ireland were turning hopefully toward the promise of a fuller, freer life in the New World. Thomas Sullivan, my mother's father, had answered the call of the western continent and awaited the coming of his daughter and son-in-law on a little farm in New Brunswick. And so, on a day in the year 1839, Patrick Nelligan and his young wife embarked from Cork on a little sailing vessel named the *Ponsila,* westward bound.

The north as well as the south of Ireland was well represented among the passengers—hot-headed, fighting disciples of Brian Boru—so the turbulent state of affairs which existed on board during the long journey can be easily imagined. The "ould sod" had hardly dropped beneath the eastern horizon—to the accompaniment of many surreptitiously shed tears—before the battle of fists and feet was in progress and it continued with but few interruptions until the sight of a long line of green in the west made the combatants forget their differences and brought them crowding to the rail with expectant faces and eager straining eyes. Fighting, to Irishmen, is not only a moral duty and a physical necessity but a positive pleasure. And when Irishmen from the two ends of the island are kept within the narrow confines of a small sailing vessel for several weeks, there is bound to be friction and much of it. Whether or not my father took part in these gay fistic combats, I do not know. Perhaps the presence of his bride deterred him. But if the grain of the chip is the grain of the block, he was among the fighters.

The fighting gave way once to something of greater interest when the *Ponsila* collided with an iceberg in a dense fog. To a heavier ship, traveling at a higher rate of speed, this would have meant disaster. But the *Ponsila* was light and slow and so did not suffer the terrible fate which was to overtake the *Titanic* many years later. Boats were lowered over the side and the sailors succeeded, with no great difficulty, in pushing the ship out of the iceberg's way with their oars.

To-day one can cross the stormy north Atlantic by boat in less than one week, but the *Ponsila* made port at Chatham, New Brunswick, only after six strenuous weeks at sea. At Chatham my father and mother were met by my grandfather, who turned over to them the little farm of about forty acres which he owned and the log cabin he had erected upon it. On this wilderness farm at Escuminac on Miramichi Bay in Northumberland County, New Brunswick, my parents began their life in the New World. And on this farm, on the thirty-first day of March, 1852, I made my unnoticeable entrance into a world in which I was to live a relatively insignificant but none the less interesting life. Three sisters, Johanna, Mary, and Catherine, listed in the order of their arrival, preceded me; one brother, Patrick, followed me.

In 1854, when I was only two years old and Pat was still a babe in arms, my father came to his death by drowning. He had traded a horse for a team of oxen at a place some distance from our farm and was returning home with the oxen. The Barnaby River had to be crossed and a ferry was used in the absence of a bridge. Frightened by the unusual and unfamiliar surroundings, the oxen became excited and unruly while crossing on the scow which served as ferry. One of them reared and knocked my father over the rail of the scow into the river. He was an excellent swimmer and could have saved himself under ordinary conditions, but he was stunned in falling when his head hit against the rail and he did not regain consciousness in time. Although only two years of age at the time, I was intensely conscious of the tragedy which had stalked into our midst and I became so sick and frightened when shown my father's remains that my life was despaired of for a week.

With the courage which has necessity for its source spurring her on, my mother continued the sturdy fight against the cold and hunger which ever lurked at the doors of such wilderness farms as ours. We all had to help in the fight and we did it as a matter of course, managing to eke out a sufficient, if slender, livelihood.

Although I do not remember many entertaining anecdotes of my boyhood days, I do retain a vivid mental picture of the frontier farm life of the time, a life now gone forever, which may prove interesting to the readers.

As I have said, we lived on a farm about forty acres in extent. Five or six cows, which the girls of the family dutifully milked every morning and evening, supplied us with milk, butter, cream, and beef. A flock of sheep which varied in size, but which was usually composed of about twenty-five head provided us with mutton and wool. The wool was woven into coarse cloth—home spun it was called although it was not made at home—and from this cloth our simple garments were made. A few hogs furnished pork which was salted and smoked. A flock of chickens supplied us with eggs and an occasional feast of fowl meat.

Wheat and potatoes were the principal crops, although other grains and vegetables were raised in quantities sufficient for our use. The wheat we took to an old-fashioned water power grist mill and returned with flour made from it after giving the miller his customary share for the grinding. From this flour our bread was made. A yeast pan, in which a small quantity was left to grow at each baking, supplied the other principal ingredient for the bread. It was sometimes kneaded for a half a day and was the lightest and tastiest bread it has ever been my lot to eat. The baking was done in an oven which could be raised or lowered from the crane in the fireplace. Live coals were placed on the top of the oven to equalize the distribution of heat.

There was a plenitude of game in the forests stretching back of the coastal farms and wild meat often occupied an important position on our table. Hunting at that time was done with the flintlock rifle, a far cry from the modern, high powered, repeating rifle of to-day. The flintlocks were heavy to carry, hard to load, difficult to aim, and often missed fire. When one missed fire it was said to have "snapped." A big Irishman weighing over two hundred pounds, who lived near us, had an unfortunate proclivity for getting drunk on Jamaica rum whenever an opportunity to do so offered itself. One night, on his way home after one of these peri-

odic sprees, he stumbled into a Scotchman's yard and fell
asleep. The Scotchman happened along, perhaps a bit
tooted, and the night being dark, he thought Pat was a bear.
He went in the house and brought out his flintlock. Taking
a careful bead on Pat's slumbering figure, he pressed the
trigger. But the gun missed fire, snapped. Sandy swore
softly as only a Scotchman can and cocked his firearm again.
Again it snapped and again Sandy swore and cocked it.
After this had happened several times, Pat, aroused from
his peaceful sleep, wrathfully raised up on one elbow and
shouted: "What in hell do you think you're snapping at?"
Whereat Sandy dropped his gun and fled, quite unnerved
by the sound of a bear using such profanity.

Fish of all kinds, especially codfish and herring, occupied
an important place in our bill of fare, as they were cheap
and easily obtainable, the ocean being at our very door. The
berries which abounded in the surrounding territory during
the summer months were picked and preserved for the long,
hard winters. Tea was the only beverage. I do not remember drinking coffee as a boy.

During the summer we all went barefoot and suffered
from painful stone bruises, but for the rest of the year we
had crude leather shoes. Side and sole leather was bought in
large pieces and made into shoes by a shoemaker who carefully took the measurements of our feet. The men wore
heavy "stogie" boots and the women's feet were clad in low
shoes of lighter material.

Our place of residence was a log cabin, and the sole
source of heat for both comfort and cooking was a huge fireplace which ate up four foot logs as if they were matches.
The exceptional ventilating qualities of this fireplace were

conducive to health, but it consumed a terrific amount of fuel, and feeding its hungry maw was one of our most important and most discouraging duties.

The settlers in that region were of French, Belgian, English, Scotch, Irish, and Portuguese origin. They were engaged in fishing and farming and got along remarkably well, there being no friction whatever between the people of different nationalities. They were all fine, common people and always ready and willing to help one another in time of need or sickness. True neighbors they were, in every sense of the word.

There was no school in the middle of winter due to the depth of the snow, but in spring and fall we trudged to the closest schoolhouse—six miles away—and absorbed what we could of the rudiments of education. The school teachers boarded from house to house, staying a certain time with each family who had children in school. They did not impress me greatly, and I remember only one definitely—a man named Walker. He had a wooden hand and that, I think, was the only reason I remember him. There were no particular grades in school, due to the fact I suppose, that there was such a small body of students and that school was such an uncertain affair. It is my remembrance that we did more fighting than studying. I was the largest in the school and, as a natural result, automatically became the school bully. I licked hell out of the rest of the students and settled their petty quarrels in a manner befitting the monarch of all he surveyed.

Every spring we youngsters were virtually forced to live on sulphur and molasses, but pleasure kept company with pain, and spring also brought one of the rarest delights for the more hardy spirits among us. The hollows in the woods filled with snow water when the spring thaws were on and we thought it the greatest sport in the world to strip and go swimming in these pools of icy water. The shock of the cold water and the tingle as it brought the blood racing to the surface was a delicious and refreshing experience. It was hardly considered conducive to health by adults at that time, but we youngsters thrived on it.

In the summer we went swimming in Miramichi Bay, the salt water of which was warm enough to bathe in. I shall never forget one harrowing experience we had. A short distance off the mainland lay Fox Island, on which berries of all kinds and particularly a kind of berry called Indian pears, grew in profusion. These Indian pears were about the size of a tame cherry, having no pit and being of a blue color. They grew on bushes or trees about eight feet in height, and I have never known them to grow elsewhere. We had gone to pick these berries in a dugout canoe made from two large pine logs spliced together. The canoe was poled in shallow water and sculled when bottom could not be reached with the pole. It was the means by which we always crossed the strait between the island and the mainland. The water was warm and inviting that day and the temptation to take a plunge was too great to withstand. Four of us older boys went in swimming and the two younger members of the party were left in the boat. We were in shallow water but the tide, which runs like a millrace and has terrific force at that place, happened to be running out, and before we were aware of what was happening, we were all swept beyond our depths. We would surely have been carried out of reach and drowned, as none of us was an expert enough swimmer to fight the tide, had it not been for the quick action of the two younger boys in the boat, who immediately sculled the craft toward us and picked us up. In so doing they did the rest of the world no great favor, perhaps, but we appreciated it at least.

The tide rises thirty feet at the mouth of the St. John River and when it goes out the vessels are left lying high and dry on the sands. I have seen teams drive around them. As is well known, the tides are unusually high all along that part of the Atlantic coast. At Calais, Maine, in 1870, the docks caught fire while the tide was out and all the vessels lying in the mud alongside the docks were burnt, because they could not be moved.

The coast was rather dangerous near where we lived and ships were occasionally wrecked off Escuminac. One English ship, a lumber schooner loaded with two-inch planks and bound for Liverpool, was considered by its captain to be unseaworthy for the long voyage ahead, so he had the pilot show him a good place to beach it and he piled it up on the sands near our home during a heavy storm. The crew was a bunch of wild Irishmen from Waterford, Ireland, and they made it a point to save all the captain's whiskey. They were brought ashore through the raging storm in a small boat and when they felt terra firma under foot again, they proceeded to celebrate the wreck of their ship by getting uproariously drunk, the captain included. The insurance company later found out about the captain purposely wrecking the boat and there was a great deal of trouble over it. But the ship was ruined and its cargo of planks was distributed all along the coast. They were picked up by the farmers and put to good use. It's an ill wind, indeed, that blows no one good.

In the year 1867, when I was only fifteen years old, I had
my first experience in the woods and my first connection with
the lumber industry, with which I was to be associated during all the rest of my life. I was no longer really needed at
home on the farm and the time had come for me, young as
I was, to strike out in the world for myself. This "striking
out" was more literal than figurative, for I "struck out" on
foot and walked more than forty miles in search of work before I found it. This was in October and the weather was
rather cold, but I suffered no particular inconvenience as I
was able to stay over night at any farmhouse along the route
without cost. Visitors were infrequent in that country at
that time, and as a result were always welcome, bringing with
them news of the world outside the narrow confines of the
wilderness farms. A crew of six men was engaged in getting out timber on the Barnaby River, a tributary of the
Miramichi River, which empties into Miramichi Bay about
thirty miles west of Escuminac lighthouse, and with them I
found work as cook. I had had no previous experience in the
occupation which I had so suddenly adopted and the ludi-
crous failures which resulted from my initial culinary efforts
can well be imagined. But experience is a good teacher,
though harsh, and necessity is the mother of invention. It
was not long before I could cook three satisfying meals a
day for a crew of six men, who were engaged in the hardest
sort of manual labor from dawn to dusk and, as a result,
were possessed of prodigious appetites.

The bill of fare had to be prepared from such staples as
bread, beans, codfish, pork, and potatoes. While these did
not admit of much variety, I was not preparing food for
jaded palates but for the healthy and voracious appetites

of hard working men, so no great variety was demanded. Good bread was the most essential article of diet, and good bread I made, in much the same manner as the bread had been made at home, baked in a sheet iron oven over an open fire. Such bread is the best and most tasty made, simply because it is baked over a slow fire in the most natural way. The beans were cooked in the famous "bean hole" fashion. There are on sale in grocery stores today, in cans, what are called "bean hole" beans, but they bear no likeness to the delicious dish cooked in a cast iron kettle in the woods. This method of preparing the beans was rather troublesome, but the result more than repaid one. A hole somewhat larger than the kettle used was dug in the ground and a good hardwood fire made in it. When the hole was well burnt-out and the sides and bottom extremely hot, the kettle containing the beans was placed on a layer of hardwood coals left in the bottom of the hole. Then live coals were piled all around the sides and on the top of the kettle which was covered with a tight fitting lid. Everything was then covered with a layer of ashes to retain the heat and the beans were left thus all night. The coals remained red hot for hours and when the beans were taken out in the morning, they were thoroughly cooked and of a flavor which could never be counterfeited in a canning factory.

Conditions in the lumber camps of New Brunswick were very crude at that time. One large room, constructed of logs, served as kitchen, dining room, and sleeping quarters. In the center of this room was a large open fireplace which served both as cook stove and heating system. A round hole in the roof of the building, directly over the fireplace, carried off most of the smoke—the rest of it dyed our hands

and faces a deeper shade of brown. I did my cooking over
the open fire and served the meals to the men on rough
benches grouped around the fireplace. We slept on beds
of brush in the same room and one huge blanket served to
keep us all warm. There were no facilities for bathing in
the camp and so no one had a bath from fall to spring, although we barbered each other and had our clothes washed
and mended at a farmhouse near by. The men washed their
hands, their feet, and their faces with snow. They were all
rather fine looking young fellows, but when they returned
to the camp after a hard day's work in the woods, hungrily
bolted their evening meal, and climbed into the crude places
of rest, they would not have appeared very prepossessing to
a delicate eye at least. But men are men, for all of that, and
these were as clean-living as conditions permitted. I look
back upon the days I spent with them with more pleasure
than otherwise.

On cold winter nights, and winter nights are always cold
in New Brunswick, the men were forced to lie exceedingly
close together in order to keep warm. They would pull the
one big blanket over them and pack themselves together as
a housewife packs and ties her spoons, back to breast, all facing in the same direction and covered with the same blanket.
When one of the jacks would become tired of lying on one
side, he would shout "Spoon!" and everyone would promptly
flop over on the other side, eventually landing in the same
compact position as before, but facing in the opposite direction.

The men were all Irish Catholics and of a rather religious
turn of mind. They faithfully repeated the Rosary during
the Lenten season and forced me to do the same, although

I couldn't quite see the importance of it. And the manner in which my cursing was curtailed was rather disagreeable. When things went a bit wrong, as things often times will, I derived a good deal of innocent pleasure, like most people, from the letting off of steam in the form of a few choice "cuss" words. But, although they did not restrain themselves in this respect, they shut down on me promptly and effectively. Perhaps they thought I was a bit young for such things.

I was supposed to be receiving a salary of fourteen dollars a month in addition to my board and bunk, but I never saw a penny of it. We were all hired with the understanding that we would be paid only if the company made money. This arrangement was accepted without question, as work was not exactly plentiful at the time. Evidently the company did not prosper—at any rate none of us was paid.

After six months in the lumber camp, from October to April, the lumbering season came to an end and I found work as a cook for a gang of English fishermen employed by the Williston Brothers, themselves Englishmen, who owned considerable equipment and had rich fishing grounds. These men were camped on Fox Island in Miramichi Bay and had their nets set in the waters close to the island. Salmon which were sold to a canning factory a couple of miles away constituted the greater part of their catch, but other fish were frequent victims of the nets, which were of the gill type. I remember watching them bring in porpoises, which have three or four inches of fat in a layer around the entire body just under the skin. The fishermen would skin the fish, cut off the fat, boil it down, and make oil of it which was used for either lighting or lubrication. During the five months

I stayed with the Williston Brothers, from May to September, I received twenty dollars per month in addition to my keep.

In October I went back into the woods again, as cook for the same crew of men I had worked for the previous winter, in the hope that I might receive some compensation for my previous winter's work through being faithful. But this hope was groundless and I received nothing for the second winter either. My duties were the same as they had been the year before with one addition. It was evidently thought that a year's experience had added to my capacity for work so, in addition to my duties as cook, I had to supply all the wood for the fireplace. This was a rather large chore as the wood had to be hauled in a sleigh or carried in from the woods and that open fire had an insatiable appetite for logs.

When the lumbering season came to an end in April, I returned home for a visit, with nothing to show mother for my eighteen months absence. That is, nothing tangible; I had gained much worthwhile experience. After a month spent at home I found work with a salmon fisherman named Henry O'Leary, who had a crew of eight or ten Frenchmen working for him at Richibucto in Kent County, New Brunswick. As before, the fishing was done with gill nets and the salmon sold to a nearby canning factory. I remained with this crew until October, increasing my now considerable amount of skill as a cook.

Chapter II   Maine and Pennsylvania

HAVING attained to the age of seventeen, the most worldly-wise age in the life of man, and with two years' experience in the world of work and men behind me, I decided that the time had come for me to strike farther afield. In October, 1869, after a brief visit at home, I took a coastwise boat from Chatham to Shediac, a small port about seventy miles south of Miramichi Bay on the New Brunswick coast. From Shediac I journeyed by rail to St. John and there took the St. John-Boston steamer as far as Eastport, Maine, where I disembarked. Eastport lies at the mouth of the St. Croix River, which is the boundary between the state of Maine and the province of New Brunswick at that point, and from there I went to Calais, Maine, twenty-five or thirty miles upstream, by riverboat.

Calais is situated on the American side of the river and St. Stephen on the Canadian side, with a toll bridge connecting the two. There was then a considerable traffic in lumber at that point, and I soon found work loading lumber boats. Maine was legally dry even at that early date, but while there was no liquor supposed to be sold, there was much which found its way down thirsty throats. There were many women plying the ancient and dishonorable trade and their best customers were the sailors employed on the boats which I was helping load. These women would come down to the docks in search of their prey and the lumber shovers, most of whom were Irish, would get me, young and full of the devil as I was, to mockingly ask them what kind of sailors they wanted. This would precipitate such an avalanche of raving, cursing, and filthy language as would turn red the ears of army mules, which are notoriously free from modesty as far as profanity is concerned. We were all highly appreciative of such coarse and robust humor as this and derived vast amusement from it.

The lumber which I helped load on the boats at Calais was sawed in the mills at Milltown, loaded on cars and railed to the Calais docks. Near Milltown was Salmon Falls, a ten or twelve foot drop in the river which was completely submerged at high tide, when the salmon could easily swim up the river to spawn. Here it was that I had my first fight.

A big bully by the name of Gillespie kicked me in the hand without any provocation, as a bunch of us were tossing pennies. He was several years older than I, but no larger, as I was then over six feet in height, and I made it plain that no one of his particular or general description could kick me in the hand and get away with it. The fight was held on the river bank near the mill site and we had a couple thousand millhands, waiting for the mill to start operating, for interested spectators. Gillespie was not quite so good as he thought he was; I proved a bit too much for him. After treating the millhands to a few minutes of interesting rough and tumble, I threw Gillespie into the river head first. His cousin then began to act nasty and kicked me in the leg with the hope of starting another fight. I thought there might be some more relatives of the wet Gillespie around, so I ignored the kick in the leg, remembering that discretion was the better part of valor, and that ended the affair. Years later in the Wisconsin woods I met men who had been among the millhand spectators and remembered me. They recalled the incident to my memory, from which it had almost been erased by subsequent encounters of the same kind.

After about a month in Calais, spent in loading lumber boats and doing any other odd jobs which presented themselves, I secured employment as cook for a crew of twenty men, who were going into the woods for the winter. This was the first time I had worked in a lumber camp in the United States and I found conditions only slightly different from what they were in New Brunswick. Nothing of particular interest happened during the winter and in the spring I returned to Calais. From there I went to Lincoln, Maine, on the Penobscot River, where I spent the summer in the employ of a farmer named William Pinkum. I had to do all sorts of farm work, and while the labor was hard, it was congenial enough and I rather enjoyed the summer I spent there. Pinkum had several boys who were young devils in every sense of the word and he hired me in the hope that I, being a bit older, might be able to curb them a bit. But his hopes were in vain and it turned out that the boys managed me much more than I managed them.

In October, 1870, I left Pinkum's farm and set out for Pennsylvania, where there was much lumbering then in progress and where, I expected, opportunity awaited me. My cousin, Dan O'Leary, joined me at Bangor and we bought tickets for Williamsport, Pennsylvania, via New York City. We were to change trains at New York City and, with characteristic greenhornishness, we got off the train the first time the conductor called out "New York City!" and found ourselves still far from our destination. Then we fell in with some professional gamblers who instantly knew us for what we were, suckers. They were big, fine-looking fellows—

Irishmen, I think, for among the members of no other race are there so many handsome rogues. Plug hats adorned their heads and they were all very well dressed. One of them, I remember, had a hand all covered with blood. My youthful imagination envisioned the possibility of that being another person's blood, but the truth was probably much more prosaic—the result of a cut from a pocketknife while whittling, or something of that sort.

These crooks explained the game of three card monte to us and, of course, it looked like a sure thing. It was. It always is. But not for the sucker. Dan was enthusiastic and felt that we had stumbled on a perfect opportunity to increase our meagre supply of cash. Our total capital was two dollars and rested in my pocket, which, perhaps, was the reason I was a bit wary of three card monte at first. But Dan's enthusiasm and the possibility of easy money carried away my doubts and I bet the two dollars, with the inevitable result. Although it was our last bit of money and meant a great deal to us, still I consider the experience one of the cheapest I ever had. It prevented me from playing the sucker with much more money in later years. I once dissuaded a large bunch of men from biting on three card monte in Menominee, Michigan, and by so doing incurred the intense displeasure of the gambler who was trying to entice them into betting their hard won cash. Having relieved us of our excess coin, the gamblers proved themselves to be very courteous gentlemen and went to much trouble in directing us to the ferry on which we crossed the Hudson River to Jersey City, where we entrained for Pennsylvania. It was fortunate for us that we had bought our tickets straight through to Williamsport. Otherwise we probably would

wo "jacks" on the river drive with their third arm, the "hook." The reader will note the stagged
nts and heavy caulk boots on the men. The cut-off, or stagged pants were a safety measure
r the logger to keep his feet and legs from becoming entangled on twigs and branches during
e dangerous work of river driving.

A picture of typical lumbermen with their high-crowned felt hats, wide suspenders and heavy wool shirts.

A huge stack of logs piled along side of the river, on the bank, awaiting the spring thaw and the drive to the mill. Note the man standing on the log in the bottom of the picture. This will give the reader some idea of the size of the log piles which were banked on the river of the pine country.

dramatic shot of the sled road through the woods. The smaller logs are piled end-wise to
e road for side loading.

Loggers are using an "A" frame to load the road sled. The "A" frame was portable, and cou be skidded or dragged to new site for additional loading.

have left our New York to Williamsport fare with the gentlemanly monte men. As it was, we hung tight to our tickets.

We arrived in Williamsport as we had left New York, penniless. We went to the proprietor of the Exchange Hotel there and, by putting up our trunks and our best clothes as security, succeeded in borrowing five dollars apiece, which was much more money in those days than it is now.

After a short stay in Williamsport, where we could find no employment we set out afoot for Westport, a little town which also lies on the west fork of the Susquehanna and some distance upstream from Williamsport. From Williamsport to Westport is an eighty mile hike. In this day and age it would be considered quite a jaunt for two fellows in search of work, but we thought nothing of it and made the journey in two days, stopping over one night at a halfway place. At Westport we hired out with a road construction crew bossed by a fine old Pennsylvania Dutchman named Jake Flackenstein. He paid us a dollar a day each and after a month there we redeemed our trunks and clothes and had them expressed to us. We did road work until the ground froze over and it was impossible to continue. The entire road crew then went into the woods along the Kettle Creek.

Along Kettle Creek, in Pennsylvania, there was some of the finest virgin timber it has ever been my privilege to see. Pine it was, almost entirely, and absolutely without rot or shake. It was almost a crime against Nature to cut it, but we lumbermen were never concerned with crimes against Nature. We heard only the demand for lumber, more lumber, and better lumber. Lumber with which to build houses, and schools, and churches, and industrial plants. Lumber with which to feed the lusty, awakening America of the last

half of the nineteenth century. Lumber with which to build
up a mechanical civilization. We heard only that demand
and we supplied it.

The method of handling the timber along the Kettle
Creek was rather unusual and interesting. A long slide,
measuring about three miles from end to end, ran down the
little valley or ravine along which we were cutting to the
Kettle Creek. This slide was constructed in the form of a
trough. One log, hewn smooth on the top, formed the bottom of the slide and two others, hewn smooth on the inside,
formed the sides. The slide rested on cross ties spaced about
four feet apart along its length. Wherever necessary, it
rested on trestlework. From this main slide similar tributaries led off like the branches of a tree or like sidings on a
main line railroad track. These tributary slides ran up the
hill or mountain sides into the woods where the cutting had
been done. In cold weather the slide was sprinkled with
water which froze into a thin and very slippery coating of
ice on which the logs slid easily and rapidly.

After being felled and stripped of branches the logs were
sawed into twelve, fourteen, and sixteen foot lengths and the
bark skinned off. They were then placed in the tributary
slides and gravity carried them down the slopes and through
the switches into the main line, where the grade was much
less steep and inertia soon overcame momentum. Teams of
horses were used to transport the logs down the main slide
to the landing on the creek. The logs rested in the trough
with the top or small ends forward so as to slide easily. Between the front end of one log and the back end of the next
ahead was about a foot of space. Each team was equipped
with a one and one half inch rope which was fastened to a

large, strong hook. The teamster would fasten the hook in the rear end of the rear log of the batch he wished to move and then cluck at his team. The rear log would start the next and the next log start the next and so on until the entire bunch would be in motion. The fact that the logs were a bit apart made it possible to haul many at a time, whereas, if the ends had been flush, inertia would have prevented the starting of more than a very few. As many as seventy-five logs were sometimes hauled down the slide by one team. When the logs occasionally jammed from slight obstructions, or at curves in the slide, the teamster broke the jam and separated the logs with a steel bar which he carried for that purpose.

The winter was very mild and the snow stood no more than one foot on the level at any time. Only during January and February was it possible to log to any extent. During the winter I spent there, we handled about six million feet of timber, but almost all of it had been cut during the previous summer. Horses were used almost entirely, there being only one team of oxen in that camp. In logging camps, as a general rule, horses were used on the long hauls and oxen on the short hauls. They were shod, of course, and it was remarkable how much even those seemingly stupid beasts could learn about their appointed tasks and the time and manner of doing them.

Conditions in this Pennsylvania camp were much improved over what they had been in New Brunswick and Maine. It was a large outfit as there were about one hundred men employed, fifty of whom worked in the woods and fifty on the landing. There were two separate camps, the men's camp or living quarters and the cook's camp or eating quarters. The food was hearty and wholesome, the beds of the best, and everything exceptionally clean.

The lumberjacks were of English, Scotch, Irish, French, and Portuguese blood. There were a few Scandinavians and still fewer Germans. I was very blonde of face and hair and, as a result, was often taken for a Scandinavian. Swedes and Norwegians often spoke to me in their native tongues before my own speech showed them their error. Everyone got along as well as might be expected and there was little trouble. The Franco-Prussian war was then in progress and occasionally there were heated arguments between the French and the Germans in the camp. There was one fine young German fellow who was absolutely docile and harmless. He had neither the inclination nor the ability to debate the rights and wrongs of two great nations engaged in conflict. Some of the Frenchmen made the most of a perfect opportunity to pick on someone without a comeback. They pestered the poor lad to distraction until finally I got sick and tired of it. I told them that if they didn't let the boy alone I'd lick the whole gang of them. Strangely enough, they took me seriously and let up on him from that time on. Not through any fear of me, I think, but because they felt a bit guilty.

On Sundays, with much leisure on our hands, we were wont to play all sorts of pranks. One of our favorite tricks was stealing pies out of the cook shanty and eating them. This, of course, did not appeal particularly to the cook and finally the cook's hired girl took the pies and hid them under her own bed. We soon discovered the hiding place and then persuaded one of our number, a half-wit named "Gundy" to go and steal them. He was caught in the act and much condemnation descended upon his witless head. The scarcity of conversational themes in camp is indicated by the fact that "Gundy's" misdemeanor provided a favorite topic for some time after. Many bright remarks were attempted and much fun poked at "Gundy" all of which he seemed to rather enjoy.

The logging season came to an end around the first of March and I went on the river with the drive. This was my first experience and it was a very edifying one. There are few kinds of labor more arduous than river driving. We got up at about three o'clock in the morning and were at work all day until darkness fell, most of the time wading in icy cold water and sometimes more than wading. Men working under such a strain as this needed stimulants. Whiskey was used and much of it. I didn't take any myself. Youth is hardy and resilient enough without whiskey.

One striking incident stands out in my memory from the many experiences I had on the Kettle Creek drive. There was much square timber got out along Kettle Creek which was taken down the river and shipped to England by vessel. The timber was bound into great square rafts and taken down stream in that way. We saw many of them drift by every day and they were a common sight, but one day we stood with bared heads while one passed. The wife of a lumberman from the State of Maine had died at some point far up stream and the river had been selected as the easiest route by which to take the body to Westport, where there were rail connections. The fine looking casket bearing the remains rested in the center of the great raft and beside it stood the sorrow-stricken husband. Great sweeps, one on the fore and the other on the aft end, each manned by three men, kept the raft in the current, which swept it downstream with the slow irresistibility of death itself. For funeral march there was only the murmur of the river among the rocks and the wind among the trees accentuating the vast silence which brooded over the wilderness. It was the most impressive funeral cortege I have ever seen. We stood with heads uncovered until a bend in the river hid the raft from sight and lumps formed in whiskey-toughened throats and tears dimmed eyes which could be—and usually were—as hard as steel.

I went down Kettle Creek and into the west fork of the Susquehanna with the drive. At Renovo, about ten miles below the mouth of Kettle Creek, we were held up for lack of sufficient water. I had caught the measles on the drive and I went back to Hammersley Forks to recuperate. Dan O'Leary was still with me and we put up at the Nelson Hotel. I somehow got a "pull" with Dora Nelson, the sister of the proprietor. Dora was an accomplished angler and caught fine brook trout regularly in the creek. Invalid as I was, she must have sympathized with me for she fed them to me regularly. But Dan couldn't get any and it peeved him greatly.

Chapter III The Great Lakes Pineries

M Y recovery from the slight attack of measles I had contracted on the drive did not take long and when I was well again I began to look toward the West with speculative eyes. The lumber industry was then at the beginning of the heyday of its prosperity in Michigan, Wisconsin, and Minnesota. It offered the greatest field of opportunity for a young fellow who had chosen logging as the business he was to follow. This was in the spring of 1871 and I was a man in all but my years. Although only nineteen, I was fully grown and measured over six feet from head to heel. I had behind me four years of experience in the woods. I was an accomplished camp cook and had worked as both woodsman and riverman. All in all, I felt that I was able to take care of myself almost anywhere, even with the three card monte men. One of my sisters was married and living in Oconto, Wisconsin, and this provided an additional reason, if I had needed any, for listening to Horace Greeley's famous advice: "Go West, young man!"

About the first of April, 1871, I set out on my long journey. From Westport I went to Erie and there boarded the train of the Lake Shore and Michigan Southern Railroad which was to carry me through to Chicago. Trains were not as fast nor as convenient in those days as they are now, but I found the journey far from tiresome, enjoying it immensely and feasting my young and eager eyes on the beautiful country through which we passed. In Chicago I had to wait a while to make connections with the Chicago and

Northwestern Railroad train which was to carry me through Milwaukee to Green Bay. Chicago was not then either as large or as wild a city as it is to-day, but it was sufficiently big so that I decided not to risk getting lost in it, and I did not venture outside the great depot during the time I was waiting for my train to start. There was a short stop at Milwaukee and there I went out of the station far enough to note that Wisconsin street was not even paved, and that there were but few indications of the great metropolis which was one day to occupy the site of the then relatively small city.

At Green Bay, which was then little more than a small country town, although one of the oldest settlements in the Northwest, my railroad journey came to an end. At that time, just before the Chicago and Northwestern Railroad pushed its line into the north, Green Bay was the railhead for all the lumbering towns as far north as the copper country. During the winter freight and supplies were hauled by teams on the ice from Green Bay to Escanaba, Manistique and other towns on the bay. In summer the rough roads which had been cut through the wilderness were used for transportation. Stage lines were well-established and stages made regular daily trips between Green Bay and the towns to the north. On one such trip a package of money sent by the express company worked loose and fell off the stage. It contained ten thousand dollars in bills. Luckily, an honest man discovered it, a man from Marinette, and returned it to the company.

To reach Oconto, which lay approximately thirty miles north along the shore of the bay, it was necessary to take passage on a flat-bottomed boat which ran between the two cities. It was a fairly large passenger craft run by steam power and owned and operated by the Hart Steamboat Company which, incidentally, ran boats in those waters until a few years ago. My long journey from Pennsylvania finally came to end on the tenth of April, 1871, in the little lumbering town of Oconto. Although rather large for a city so far north—its inhabitants then numbered four thousand—it was as typical a lumbering town as you could ever have encountered. Plank sidewalks, punctured and chipped by the calked boots of rivermen, lined the muddy streets, and behind the sidewalks stood the mercantile establishments of the booming little city, most of them saloons engaged in the lucrative business of extracting the "filthy lucre" from the "two-way" pockets of spendthrift lumberjacks.

The lumberjacks of the Northwest, those hardy forerunners of our present-day civilization, walked the streets in all their pristine glory. I shall never forget how splendid those young giants of the North first appeared to my impressionable eyes. They were strong and wild in both body and spirit, with the careless masculine beauty of men who live free lives in the open air. They seemed the finest specimens of manhood I had ever seen. They were magnificent; even in their annual periods of dissipation, when they flung away the wages of a winter's work in a wild orgy lasting only a week or so, they were magnificent. Drunk or sober, they would fight at the drop of a hat and fight to a bitter finish. They had their code and it was a chivalrous code. Rough in dress and speech and manners, gaining their livelihood by the hardest kind of manual labor, living, loving, and laughing crudely, still they were gentlemen. No man could offend, insult, or molest a woman on the street, no man could even speak lightly of a woman of good reputation without suffering swift and violent justice at the hands of his fellows. So here I pay my tribute to the lumberjack of the Middle West, an unsung pioneer, the hero of a passing epic drama, a gentleman and—more than a gentleman—a man! May the memory of his days and ways endure!

About a month after my arrival in Oconto, I went to work on the north branch of the Oconto River for a company owned by the Sargent Brothers and William Bransfield, who was my brother-in-law. We put in about two months there and arrived back in Oconto on a Monday in about the middle of June. The town was still in a ferment over a tragedy which had happened the previous evening, an occurrence which illustrates how easy it was to arouse mob spirit in those turbulent days and what terrible injustices often resulted.

Two men known as the Klause Brothers ran a saloon in the city and were in the habit of holding dances every Sunday evening in the dance hall over the saloon. They attempted to maintain the place in a respectable way and to keep disorder at as low an ebb as was possible under the circumstances. Oconto, like every other lumbering town of its size in that day, had a gang of rowdies who were "hard" in every sense of the word. It had become their habit to drop in on the Sunday night dances and raise particular hell. Finally the owners got tired of this lawless tyranny exercised by a gang and determined to bring it to an end if possible. So they posted an armed "bouncer" at the door with orders to keep the rowdies out at any cost. On the Sunday evening of which I speak, the dance was in full progress when the gang of roughnecks, as usual, appeared on the scene with mischief in their manner. Denny White, who came from Chatham on Miramichi Bay near my home in New Brunswick, was the instigator of all the trouble. They started to "clean up" the place and the "bouncer" becoming unduly excited, pulled his revolver, aimed at one of the disturbers of the peace, and fired. But his arm was shaky and, instead of hitting the man he intended to, he shot an innocent bystander, a young fellow named Joseph Rule. The affair came to a head so suddenly and the outcome was so tragic that the "bouncer" was taken into custody and lodged in the town jail before mob spirit could be sufficiently aroused to prevent it.

Unfortunately for the doorkeeper, the young man who had been hit by his ill aimed bullet, Joseph Rule, was one of the most popular young fellows in the town. He had, of course, been guilty of no wrong whatever and the townspeople were bent on bloody revenge. They forgot that he was the victim of a tragic mischance and not of a murder; they forgot that the doorkeeper had really been representing the forces of law and order; they forgot everything except that vengeance, in the form of a manila necktie, must strike the helpless "bouncer."

A crowd gathered around the little jail and mob spirit was rapidly aroused to a point of action. The leaders obtained a small saw log and, using it for a battering ram, knocked in the door of the jail. The unlucky poor shot was dragged from his cell and out of the jail, screaming and protesting piteously. When they reached the middle of the bridge across the river, the victim, who had by then given up all hope of his life, begged and pleaded for time to say his prayers. The request was brutally denied, the lynchers

probably fearing that the sound of this poor doomed creature's prayers might arouse a bit of human mercy in their
hearts. They dragged him on, across the bridge and to the
spot where, strangely enough, a house of justice now stands,
the Oconto court house. There they hung him to a tree and
Joseph Rule was avenged, if such be vengeance.

One cannot help but speculate on the erection of a court
house on that spot. One wonders whether it is a symbol of
justice rising over injustice, or whether it is built upon foundations of injustice? Time itself has avenged the death of
the lynching victim. One of his murderers later hung himself. Others suffered violent deaths and all of the active participants in that sordid, gruesome, and disgraceful affair encountered consistent misfortune.

A similar affair occurred nine years later, in 1880, at Menominee, Michigan. Two brothers named McDonald, who
hailed from some place in Canada, had got themselves into
difficulties with the county sheriff and each had received a
year in Jackson prison. After serving their sentences they
returned to Menominee and shortly found themselves in a
desperate fight with some half-breed Indians at a notorious
hangout, famous for its rough frequenters. In the heat of
the encounter one of the McDonald boys slid the blade of a
pocketknife between the ribs of one of the Indians and killed
him. The two were seized by the authorities, put under arrest, and placed in jail.

Here, as at Oconto, some of the townspeople became a
bit bloodthirsty and decided to take justice into their own
hands. A mob numbering between five hundred and a thousand people smashed in the doors of the jail and seized the
luckless McDonald boys. Nooses were placed around their

necks and they were mercilessly dragged about a mile out of town and hung to a tree. They never knew they were hung. They were dead, their necks broken and the air cut off from their tortured lungs, long before the ropes were thrown over limbs and their bodies left to dangle in the wind, gruesome testimony to the brutishness of man. Most of the leaders in that mob were drunk and some were so far removed from any semblance of humanity that they rode on the backs of the victims as they were cruelly dragged, face down, along the rough road.

In this case, as in the Oconto affair, all the principal participants seemed to suffer consistent ill-luck afterwards. Every business man who was in the mob lost his money, his business, and the respect in which he was held by the public. It was a pitiful and disgusting exhibition of beastliness. What made it the more regrettable was that the McDonald boys could easily have been saved had the sheriff been any good.

I celebrated the Fourth of July, 1871, in Oconto and shortly afterward secured employment with Anson Eldred and his son Howard at Stiles, a little lumbering town on the Oconto River between Oconto Falls and Oconto. The Eldreds were large, efficient operators. Anson Eldred was an exceptionally brilliant man, honest but tyrannical. We were doing summer logging and he used the most modern methods of handling logs known at that time, methods which would, of course, appear rather crude to-day. After the logs were cut into lumber, the lumber was gathered in rafts and floated down the Oconto River to Green Bay, where it was loaded on vessels bound for Chicago and lake ports in the east.

I continued in the employ of the Eldreds throughout the summer, and during either the latter part of August or the first part of September was hired by Sargent and Bransfield to go up into the total wilderness along the north branch of the Oconto River. There I was stationed in charge of about twenty head of oxen, which were enjoying a life of leisure until winter should come and the camps open for operations, when they would go under the yokes once more. I had a tent and a light camping outfit and my job was easy, as I had only to keep general track of the oxen and see that they did not go astray and that they were not shot by anyone.

After a little more than a month in the woods, I ran out of supplies and on Sunday, the eighth of October, 1871, a memorable date in the history of the Northwest, I set out for a supply depot or trading post about twelve miles distant with the object of replenishing my depleted larder. Twelve miles was but a short hike for me in those days and I had reached my destination, bought my supplies, filled my pack, and was on my way back by early afternoon. The air was hazy with smoke and it was evident that the demon, forest fire, was riding the wilderness. When still about four miles from camp my path was blocked by fire and I was forced to change direction and head for a small farm owned by Anson Eldred which lay some distance from my original destination. With good fortune I was able to reach the farm in safety, and I found the man in charge, an excitable Frenchman, hysterical with fear and sobbing out his belief that the place would be burned to the ground. In truth, there were sufficient grounds for his fears for the forest seemed afire in all directions. But I was not accustomed to lying down on any job, however impossible it might seem to be. So, I hitched a team of oxen and with the help of the Frenchman who soon forgot his fears and tears under pressure of hard work, we hauled barrels of water on a sort of drag and so were able to fight back the vanguards of fire, which were eating away at stumps and trees in the little clearing around the house. We worked steadily all through the night and by dawn the flames were under control and the farm buildings saved.

Forest fires all over the Northwest reached their climax between that night and the tenth, although they had been burning for days before and continued for days after in some places. They ravaged the lower as well as the upper peninsula of Michigan and left death and destruction in their wake. Over one thousand persons perished and three thousand were left homeless and destitute by the flames. In Wisconsin the tornado of fire reached its peak of destruction at Peshtigo, but raged for miles northward before it died out. The great Chicago fire, which is mistakenly supposed to have been started because of the indiscretion of a certain Mrs. O'Leary's cow in kicking over the lantern by the light of which she was being milked, started on that fatal night. I have heard that the Chicago fire chief testified to the fact that many fires broke out stimultaneously in all parts of the city. That would agree with the testimony of many witnesses through the Northwest that there was "fire in the air." Many explanations have been offered, among them being the one that the fires resulted from the tail of a comet which passed close to the earth. A more reasonable and logical explanation is that these fires resulted from spontaneous combustion after a period of excessive dryness had paved the way. Many witnesses claim to have seen balls of fire in the air, which have been explained as being accumulations of combustible gas.

Many worthy accounts have been written of the Peshtigo fire and I shall make no attempt to improve on them here, as my knowledge of it is based simply upon what I have heard and read of it. I shall, however, give a brief general account of that terrible disaster. The lumbering town of Peshtigo had then a population of about fifteen hundred people and was located on the river from which it takes its name, some miles north of Oconto. Of that population almost one half, or seven hundred, died from burns or suffocation. The inhabitants were panic-stricken and sought refuge in all directions, often without the slightest reason. Many of them crowded into the boarding house under the false impression that it would survive and suffered fiery deaths when it collapsed like paper under the onslaught of flames. Others, with a bit more reason, sought safety by the river. Some huddled together on the bridge. As the heat increased some horses crowded on to the shaky structure and down it went to destruction, horses, people, and timbers, in a hellish mixture. Hundreds of people had hair burnt from their heads. There was a general impression that the world had come to an end and in the villages surrounding Peshtigo many people committed suicide, evidently desirous of being among the first to greet Gabriel. In the wake of the tragedy came wonderful relief work and funds and supplies were received from all over the country. Tents were set up and reconstruction began. Thieving was, of course, rampant. The much talked of "ill wind that blows no one good" was not the wind that fanned the flames at Peshtigo, for that particu-

lar wind blew a few rogues into the logging business with
supplies which they stole.

Comedy strode hand in hand with tragedy and there were
many incidents which proved laughable after the pain had
subsided. One of these incidents I shall relate, mainly because it deals with a man to whose unusual career an entire
chapter will be devoted later.* This man I shall call Ferguson, because it is so unlike his real name.

There were few banks in that region in those days and
people who dealt much in money were forced to carry considerable currency on their persons. One such man, a col-
lector, made an attempt to flee from Peshtigo, as did many
others, and was overcome by the heat and smoke. Dropping
to the road, he died; the flames crept around him and burned
him so that blood trickled from his side and stained a huge
bunch of bills, three or four thousand dollars, sticking out
from his pocket. The fire died, morning came and with it
the dead man's brother and a group of morbid onlookers.
Among them was Ferguson, the man of whom we have
spoken. The brother was overcome with grief, and with
tear-wet eyes looked down upon the body of the dead man.
Ferguson also appeared to be grief-stricken, but his keen
eyes were undimmed by tears and they spotted the pack of
blood stained bills. Then, indeed, the hypocritical rogue was
overcome with sorrow. He kneeled by the dead man's side,
took the bunch of currency—a blood stain or so was a small
matter in Ferguson's estimation—and, in a sorrow-shaken
voice, announced that he would keep these as a remembrance
of the dead man. The bluff was so bold that it almost
worked. But the brother awoke to his senses in time and,

*The "chapter" has been omitted. Editor.

taking the bills, reminded Ferguson that the newspaper by the side of the deceased would perhaps do as well for a remembrance. With the greatest alacrity, for he was always quick to realize defeat, Ferguson agreed and, taking the paper, departed, wishing no doubt that he had encountered such a lucrative opportunity under cover of darkness.

From Peshtigo the tornado of fire tore its way northward to within one mile of Marinette where, for some unknown reason, it died out on the very outskirts of the city. It was generally believed in Marinette that the city would be destroyed and large numbers of people were put on great lumber scows and towed to a safe distance out on Green Bay by tug boats.

I worked for Sargent and Bransfield on Little River, a tributary of the Oconto River, during the winter of 1871-72 and there, although it was some distance from Peshtigo, where the fire reached its peak, the flames had left their terrible marks. I was employed as a swamper that season and was as black as an Ethiopian all winter long from handling burnt wood. Stumps had been reduced to ashes, and we saved much good pine timber which was being destroyed by the worms which infest the trees after fire. I did not see a green tree during the entire winter. We often ran across the carcasses of animals which had met their ends in the inferno. They had been caught in a veritable trap of fire and consumed. I remember seeing the bodies of deer with their legs burnt off. Many animals, of course, were able to save themselves by avoiding the path of the flames but, in the territory through which the fire swept, I think there were more lost than saved.

In 1872 I witnessed the results of another of Nature's playful, prankish moods when the little village of Pensaukee was blown all to pieces by a tornado. The storm struck the village about seven o'clock in the evening and demolished everything in its path. Lumber scows were upset in the river; the railroad bridge was blown out of line; a fine new summer hotel was stripped clean of brick veneer from foundation to roof; and a sawing machine was blown through a cow, typical of the ghastly but humorous tricks tornadoes love to play. No people were killed, but several were hurt. I was on the train bound from Oconto to Green Bay when it happened and the train was run back to Oconto for medical assistance.

The hotel manager's daughter and the station agent were seated on the hotel porch when the wind struck with characteristic and violent unexpectedness. They were picked up and dropped forty yards away on the edge of the lake. The girl was unhurt but the man's hand was slightly cut by flying glass. The station agent had a head shaped like a five cent watermelon and a nose which bore a great resemblance to the jib sail of a yacht. People claimed that this head and nose changed the direction of the wind when he turned around and that was all that saved the pair of them. They were married shortly after the tornado, probably feeling that any storms on the sea of matrimony would be mere zephyrs compared with the wind they had weathered together.

Chapter IV Logging and Land Looking

THE Chicago and Northwestern Railroad extended their road northward from Green Bay in the summer of 1871 and proceeded as far as Menominee, Michigan, before winter put a stop to construction. In the spring of 1872 the work was continued and the railroad finally completed to Escanaba, Michigan. For this pioneer venture in railroad building the company received extensive land grants in upper Wisconsin and Michigan from the federal government, grants of such value that it was able to pay many times over the cost of construction of the road. It also had, just as any individual, the right to enter any lands it chose at the government land office at a price of one dollar and a quarter an acre.

Shortly after the completion of the road to Escanaba the officials of the company decided that it might be a profitable investment to enter some of the lands in the upper peninsula of Michigan, knowing that the region was rich in timber and might have extensive mineral deposits. So, in the spring of 1873, the Northwestern sent out a crew of forty men to cruise lands in the upper peninsula of Michigan. I was one of the forty. On the twelfth day of May we started out from the little mining village of Republic, Michigan. We were divided into eight crews of five men each and each crew had a flat bottomed boat called a bateau in which it made its way down the Michigamme River, along which we were cruising the land. These boats were large enough to carry five men and their supplies and equipment and they were heavy enough so that it took four men to carry one over the portages. There was a supply depot at the mouth of the Deer River, where it empties into the Michigamme, and when we ran low on supplies we would make a trip to this depot and replenish our larder. The eight crews ran their bateaux to various predetermined points along the Michigamme and then each crew would hike to the district it had to look over, usually taking with it enough supplies to last for a period of two or three weeks. A sack of flour weighing one hundred pounds would be strapped to one man's back; a hundred weight of meat would be slung from another man's shoulders; and the other three members of the crew would carry about one hundred pounds each of mixed supplies and equipment. Thus heavily weighted down, we would make our way to the township we wished to cruise and make camp until that particular part of the job was finished, or until we had run out of supplies. The mosquitoes and sandflies were almost unbearable during the months of May, June, and July, but after that they eased up a bit and life became endurable.

It was our duty to locate and estimate pine timber and also to note any indications of iron ore. The business of land looking or timber cruising is rather interesting and I shall here attempt to give a brief description of the manner in which it is done. The first thing to be done is to locate the stakes and markings of the government survey and to find out the variation on which the north and south, or longitudinal, lines are running. If one has the field notes of the government survey one does not have to make use of the transit compass to determine this variation. After having located one of the government stakes, an experienced cruiser easily finds the rest. He takes a hand compass and strikes off on foot, counting his paces: five hundred paces for eighty rods, one thousand paces for one half mile, two thousand paces for one mile. Eighteen hundred paces will cover a mile in good, open hardwood country. The sections are usually fractional, having either more or less than six hundred and forty acres, due to the variation from parallel of the longitudinal lines. Government maps show how many acres the section has more or less than six hundred and forty. Lakes of any size are deducted from the section's acreage.

Estimating the timber in a given section is the next operation. A tract about two hundred feet square, which appears to be typical of the rest of the section, is paced off. This two hundred feet square tract is approximately an acre. The number of trees on it is counted or estimated and then a few of them scaled to get an average of the number of board feet to the tree. To determine the amount of merchantable timber in a section, the average number of board feet to the tree is multiplied by the approximate number of trees on the acre, and the total is then multiplied by six hundred and forty, the number of acres in a section. A good timber estimator must have keen judgment and wide experience in the field, for timber is very deceptive. Much of it is fine looking from the outside, but when it is cut one finds that it is simply sap and bark, probably because the tops have been broken off and the rains have seeped in and rotted it. The quality of timber is easily determined when it is felled. If it is good it will break off six or eight inches at the butt due to the weight at the top, but if it is rotten the entire butt must be sawed through before it falls due to the lightness of the top.

With much practice, a cruiser becomes marvelously ac-
curate in estimating and can judge distances, areas, and
timber footage without going through the rigmarole I have
described. We often used to climb a tall tree—with the
help of a ladder made from a small tree by trimming off
the branches and leaving the spikes on—and from a point
of vantage in the top of such a tree, usually on high ground,
were able to locate and estimate timber in a large surrounding area with a fair degree of accuracy. We would take
our compass and maps with us and so conduct a sort of
aerial survey of the region within sight of our eyrie.

Our crew of five was made up of Peter Jameson, John
Archibald, James Sargent, Tom DeWire, and myself. Jameson, Archibald, and Sargent were all good land lookers, but
DeWire was a no-good tame ape who was always seeing
wild animals and getting himself in trouble. One day he
got lost in the woods a short distance from camp and instead
of trying to keep his head and find his way out, he got the
idea a mountain lion was chasing him and he began to yell.
Jameson was in camp and the shouting scared him stiff. He
started running into the brush, going like the whirlwinds of
hades. I saw him and went after him. Fortunately for
Jameson, my legs were a bit longer and faster than his and
I soon caught him.

"What in hell's the matter, Pete?" I asked him.

"The Indians have got Tom," he panted.

"I hope to [————] they keep him!" I answered.

But no such luck. I towed Pete back to camp and got
him quieted down a bit when along came DeWire, the damn
fool, looking for the butcher knife so that he could fight a
mountain lion which had been chasing him.

"You've lost your head, if you ever had one!" I told him. But that particular part of his anatomy was as hard as it was empty and he took no offense. If Jameson had got away from me and into the woods, he would surely have been lost and would have had great difficulty in making his way out again as he had no compass with him. Jameson's nerves were all shot, as well they might have been. Sargent and Archibald came into camp a couple of hours later and that ended that episode. It is the height of folly for a man to go into the woods entirely alone, even with the necessary equipment for finding his way and camping out. One is too apt to fall and break a limb, or meet with some other mishap, and the wilderness is merciless to those it catches in its toils.

Winter time is by far the best time of year to cruise your timber and it is not so much of a hardship to camp out at that time of year as it seems at first thought, in spite of the fact that the weather is extremely cold, the ground frozen, and the snow two or three feet deep on the level. We usually camped on high land where it was possible to obtain plenty of hardwood for campfires, maple, if possible. After picking the site for our camp, we would make big scoops out of wood—rough snow-shovels—and clear the snow off a large space of ground. Then a great log fire would be built on the ground over which we intended to pitch our tent and this fire would thaw out, dry, and warm the ground. After clearing away the remains of the fire we would heap balsam or hemlock boughs on the space and pile them together with the butts on the ground and the tips pointing toward the head of the bed. This made a fine mattress, springy and comfortable, on which to sleep. The next thing to do was to pitch the tent, which was shed style and was placed as near as possible to the fire without burning it. Over the tent we built a second roof of poles and a layer of evergreen boughs, leaving an air space between the two which insulated us against the cold and protected us from frost. The fire was kept going all night and used about two cords of wood every twenty-four hours.

As far as the actual moving of equipment was concerned, it was easier to change camp in winter than in summer. Our camp outfit was packed and placed on a toboggan and we would tramp a path ahead of it with our snowshoes, a path along which the toboggan slid smoothly and easily. The best way to camp out in winter is to have a house tent and a small sheet-iron stove with a pipe which passes through a hole in the roof well-secured with tin. Such equipment will save much time and labor otherwise spent in cutting tremendous quantities of wood and in preparing the shed tent. Perhaps the worst feature of timber cruising in winter is the running of the hand compass with one's bare hands. It is work that cannot be done with mittens or gloves on.

One of the great advantages of cruising in the winter is that one is able to cross the lakes on the ice and can travel through the woods on snowshoes. But there is great danger in trusting the ice of lakes and many a woodsman has paid the supreme price for his carelessness. Drownings in ice-covered lakes were common tragedies in early days. A land looker from Oconto, Wisconsin, named Samuel Orr, was up in Minnesota looking timber one year and lost his life in that manner. He slept at a logging camp one night and the next morning set out across the lake. The ice was too thin to hold him up and he broke through. When he saw that he couldn't save himself he took off his hat and threw it on the ice, where it was found later, the wilderness' receipt for another victim.

The getting of water for drinking and cooking while out land looking was no problem in winter, for there was always snow to melt, but it presented real difficulties, sometimes, in the summer months. Once when Flannigan, who was my partner, and myself were cruising some timber on the headwaters of the Ford River we couldn't find any water for two days and the weather was uncomfortably warm. Thirst, we found, was very disagreeable, much more so than hunger. On the third day we ran across a mud lake, the water of which we had to boil before using. We learned a bitter lesson and on our next trip into the woods we carried two canteens. Most of the small streams were dry so that was the only safe way.

After six hard months of land looking for the Northwestern, we finished our job on the upper Wolf River and from there I returned to Oconto. There were six or seven inches of snow on the level when we quit work about the first of November and came out of the woods. Our clothes were in rags and our shoes and socks were worn out, so that our feet were exposed to the snow and cold. But we didn't stay that way long, for each of us had a fairly sizable stake. We received $3.50 per day for the time we were in the woods and the company supplied our provisions and equipment.

The Chicago and Northwestern entered only a small part of the lands we cruised, this being due to the panic of 1873. The man who made out the reports on the land entered it and later made loads of money out of it.

Although my six months of land looking in the Michigan wilderness had been a trying and tiring experience, it had so hardened me that I was ready for work again after a short period of rest and I went into the woods for the winter. Following the spring drive of 1874, I went to Green Bay and enrolled in the Commercial Business College, which was under the able direction of Mr. A. C. Blackman. There was nothing in particular for a woodsman to do during the summer months and as I did not look favorably, as do most lumberjacks, on an extended period of carousing and loafing, I thought it was an opportune time to improve my sadly neglected education and learn a bit more than I knew about the methods of transacting business. I was now fairly well-versed in the practical side of the logging business, but I needed more education in business methods. In my case the usual procedure was reversed, and I received my experience before my "book-larnin'." Not all lumberjacks were wastrels and there were quite a few who enrolled in the business college at the same time I did. Several of them are independently wealthy today, partly due, no doubt, to the education they picked up there.

In the fall I left Green Bay and returned to Oconto. There I hired out to Henry Sargent to work in the pine woods on the main Oconto River, some fifty miles west of the city. On the first of November, 1874, we started out on the tote road along the river with a cavalcade of two four-horse teams and two yokes of oxen hauling wagons loaded with the necessary supplies and equipment to open a logging camp in the woods. When we arrived at the river crossing we found that the stream was frozen over with a coating of ice several inches thick, but not so thick that we considered it safe for the oxen and horses to cross on. So we sawed a passage in the ice about six feet in width and crowded all the cut ice out of the way. Along the edges of this cut we laid twelve inch boards about sixteen feet in length, end to end, so that the weight of a man walking along the edge was distributed and there was no danger of his falling through. The four oxen were unhitched and two lines tied to each one's head. Two men, one on each side of the passageway, held the lines as they walked across and, with its head thus held above water, each ox safely swam the chilling stream. When we got them all across the river we dried them carefully and rubbed them down briskly, after which they were covered with blankets and fed. The camp equipage was carried across the stream by hand and we carried the wagon boxes over after uncoupling them from the wagons, which were then pulled across. The horse teams returned to Oconto with their empty wagons for more supplies. The weather continued to be quite mild until after New Year's Day when we had a cold spell and the river froze solid enough so it was safe for teams to cross. Until that time we were forced to transfer all supplies and equipment at the river.

We arrived at the location on which we were to build our camp that afternoon, the third day out from Oconto, and immediately set to work. Part of the crew was put to work at the hurry-up job of preparing a shelter of sorts for the crew to use that night and the nights to follow until we completed construction of the camp. The balance began cleaning the ground for the camp and barns. The next morning we started felling timber with which to build the main camp structure. It was, necessarily, rather a large structure and the logs we felled for it were mostly sixty feet in length, the length of the building. It was about half that in width and, of course, only one story high. A partition divided the men's quarters, in which the men gathered and slept, from the cook's quarters or the kitchen and eating room. About a week sufficed to complete construction work and then we commenced cutting roads and preparing for the winter's harvest of timber.

The season came to an end in April, 1875, in the camps, but I was on the river until June with the drive. When the logs were safely stowed away in the booms, ready to enter the maws of the hungry mills, I returned to Green Bay and again enrolled in the business college there, where I spent the summer. About the first of October I entered the employ of Crawford and McKillop, who were engaged in logging operations on the Menominee and Michigamme rivers. Along with the members of the crew with which I had hired out, I made another eighty mile hike in two days with one stopover, as hard a hike as Dan O'Leary and I made from Williamsport to Westport, in Pennsylvania. We walked that distance up the Menominee River to a place called Badwater, an Indian village, and there we went to work cutting a supply road eight miles through the woods to the Michigamme River. After completing the supply road, we went to work in the camps for the winter. That same winter, 1875-76, the Chicago and Northwestern commenced construction on its Menominee River branch, which was built from Powers, Michigan, to Iron Mountain, Michigan. This was much appreciated by the lumber companies as it made it possible for them to get their men, equipment, and supplies close to the scenes of their operations along the upper tributaries of the Menominee with ease. When that branch was completed, it was no longer necessary for the men to make such hikes as we made that fall. I continued in the employ of Crawford and McKillop until the spring of 1877.

When the camps closed for the season early that spring, I went in to Oconto and from there to Green Bay to have some clothes made by Mr. Hoffman, a merchant tailor there. While in Green Bay I stayed at the Adams House, a first-rate hostelry managed by a Mr. Allen. Near the hotel there was a saloon owned and operated by Jack Brennan. It was a hard joint—a loafing place for gamblers, third-rate prize fighters, and the rest of the tough element. Like most managers of such places, Brennan had what was known as a "bunco-steerer" to lure strangers into the dive. This "bunco-steerer" was a fine looking, agreeable young chap, a German, and I, all unaware of his identity, soon struck up an acquaintance with him around the hotel. He probably thought that, as I was a lumberjack just out of the woods, I was a bird all ready for plucking. He urged me to take a walk over to Brennan's and look the place over and I finally agreed. As usual, there was a crowd of toughs there and they welcomed me like a long lost brother.

"Have a drink!" they invited.

"Thanks, but I don't drink!" I replied.

"Well, let's smoke," they urged. "Have a cigar."

"No, thanks! I don't smoke," I declined.

"Then let's play some cards!"

"Sorry, but I don't gamble either!"

I could see that I was in a pretty tight place. They were hungry for prey and I was apt to get a blackjack in the head any moment. So, when I was ready to go, which was soon,

I kept my face toward them and my back toward the door and backed out.

About six months later I was at the Oconto depot one day when a train from the north pulled in and my friend, the "bunco-steerer" hopped off. I caught his eye and told him I wanted to see him a minute. He didn't remember me, so he readily agreed to go around to the other side of the depot.

"Do you know me?" I asked him when we were out of sight of other people.

"No, I don't," he replied.

"Don't you remember the time you steered me into Jack Brennan's dive in Green Bay?" I asked him.

He said, "No!" again, but his eyes belied his word. I caught him by the ears and drove his head back against the wall of the depot with a good deal of force. That broke his bluff and he began to beg off, saying he was paid for that kind of work. Although that was a rather flimsy excuse, it was the truth and I let him go unharmed after throwing a good scare into him. Thereafter, he was probably a bit more careful about picking his victims.

Jack Brennan located in Marinette some time afterward. He opened a big saloon and enjoyed the lumberjacks' trade for years. When the men piled off the train after a winter in the woods, thirsty for wine, women, and song, Jack would be standing outside his resort with a spotless white apron on him and a jovial smile of good-fellowship wreathing his sinful old features. He would invite them into his saloon and treat them to beer and whiskey to get them started. His saloon was a real resort, catering to every desire of the woods-weary lumberjacks. There was a lunch counter and a vaudeville performance. Music was going continually and

there were women to dance with. The lumberjack was
royally entertained according to his own tastes until his last
dollar had been extracted from him and then he was tossed
aside like a dirty shirt. Men of Jack Brennan's stamp knew
that the gutter was the only proper place for a penniless
lumberjack. The strangest part of it all was that but few
of the saloonkeepers who made so much money were able to
keep any of it. Brennan's business finally went to pieces
and he left Marinette without a dollar to his name.

There was always some time between the closing of the
camps and the spring drive and in April, 1877, after the incident related, I returned to Oconto and hired out to Anson
Eldred and his son Howard to help drive their saw logs out
of the north branch of the Oconto River. The company
that was driving logs behind us on the stream had some
trouble with its men and about thirty of them left the works.
The same number of Indians were hired in their places.

It wasn't long before trouble broke, for Indians and
white men seldom got along very well on the drive. About
seven of us were bringing up the rear of our drive and working just below one of the dams when twenty or more of the
Indians congregated on the dam and began jeering and
yelling at us. We paid no attention to them and went on
with our work until we were through on one side of the
stream and had to cross to the other side. When we arrived
at the dam the Indians were still there, still shouting insults
at us.

We didn't waste any time with words, but went after
them with fist and foot. About half of them we knocked
into the dam pond, from which they emerged dripping and
spiritless. A few of them fell over the dam into the river

below. They all crawled out on the shore and started to run downstream, with us hot on their trail. We chased them only far enough to give them a good scare, but they kept on running for eight miles, all the way to the Waupee.

Mose Thompson, woods superintendent for the Oconto Company, was going up the river with his team and when the horses saw that gang of Indians coming down the road toward them, they took to the woods. It required all of Mose's skill to get them back on the road again. After the affair at the dam, we had trouble continually with the members of the other crew and when both outfits got into Stiles the crisis came.

As I didn't smoke or drink or carouse to any extent, I kept away from the dance and roughhouse with which the river drivers celebrated their arrival at Stiles. Along toward morning the bully of the other gang—a white man—made boastfully brave by drink, sent me word that he could lick any man who would knock down an Indian who was working with him. I received the message but wasn't much disturbed by it. In an hour or two morning dawned. Half a dozen of us ate breakfast and were ready to leave for the rear of the drive—about two miles upstream from Stiles. Before leaving we stepped into the bar of the Forest House. The bully was there—still inspired by liquid courage. He swaggered up to me.

"Did you get the word I sent you?" he asked.

"Yes, I did. Can you do it?" I came back.

"You bet I can!" he replied.

So, without further words, we stepped up and got to work. It must have been a good fight to watch, but it was a better fight to fight. We kept at it for half an hour without an instant's let up. I knocked him down several times, tore all the clothing off his back—and he had plenty on—and kept at him until he couldn't move. He wasn't exactly a sweet looking sight when I got through with him and I couldn't have had such a prepossessing appearance myself. Marquis of Queensbury rules had no place in a rough and tumble lumberjack fight. That encounter ended the trouble on the drive. I was never able to find out just why the fellow wanted to lick me. The other men were just as much implicated in chasing the Indians as I was.

The Indians, of course, were government charges and the Indian agent tried to make trouble for us for interfering with them. For a while we had a merry prospect of a trip to Leavenworth staring us in the face, but Mr. Eldred used his influence and the matter was patched up and glossed over. The sight of those Indians running pell mell away from us was almost worth paying for with a couple of years in Leavenworth, anyway. Leavenworth couldn't have been much more strenuous than river driving—working in icy water from four o'clock in the morning until eight in the evening.

As I mentioned before, trouble almost always broke out when there were crews of whites and Indians on the same stream. I remember hearing of a famous fight which occurred on the Oconto River some years before the Civil War. There was a crew of Indians on one side of the river and a crew of white men on the other side—both crews numbering about the same. Trouble brewed between the two outfits all the way downstream, but the climax didn't come until the drive hit Oconto Falls. There it was decided that things had to be settled with a fight, so the Indians selected as their champion a big half-breed named John Galineau and the white crew picked a strapping young Irishman named **Pat Golden** to represent them. The cook didn't want to see murder done, so he had all the carving knives thrown in the river. Golden and Galineau were pretty well matched and the fight lasted for half an hour with honors about even. Then Golden was lucky enough to land a good punch on the half-breed's jaw. It sent him staggering backwards until he tipped and fell into a large boiler of hot bean soup. That settled the trouble and the Indians, fearful of more such treatment, legged it for the woods as fast as they could go.

Chapter V Cone to Consumer

SOME years ago, I have been told, a tramp stopped at one of the Sawyer-Goodman Company's logging camps and, in the immemorial manner of vagabonds, asked for something to eat.

"How'd you like to go to work?" asked the foreman, more as a matter of form than anything else, although he was in need of men.

"Fine!" replied the hobo, shocking everybody within hearing by such a drastic violation of the ancient and honorable tradition of indolence among tramps.

"What can you do?" queried the foreman when he had recovered from the first shock of surprise.

"Sky-hook!" answered the tramp.

"All right," said the foreman, "Grab one of those cant-hooks and mount that log car."

The hobo did as he was told and proved to be a good workman, but shortly had the misfortune to be the victim of an accident in which one of his legs was fractured. "Top-loading," or "sky-hooking," is dangerous work. The company sent him to St. Joseph's Hospital in Menominee, Michigan. During an extended convalescence he got into a conversation with one of the nuns and she asked him how he had suffered his injury.

"Well, sister," said the hobo-jack. "It was this way. I dropped in at one of the Sawyer-Goodman Company's camps and as I was the first 'gazebo' who came down the 'pike' and the 'push' needed men, he put me to work 'sky-hooking' or 'top-loading.' The first thing the 'ground-hog' did was to send me up a 'blue.' I hollered at him to throw a 'Saginaw' into her, but he 'Saint Croixed' her. Then he 'gummed' her and the result was I got my 'stem' cracked. See?"

"Dear me!" said the nurse, "I haven't the least idea what you're talking about."

"By God, sister," replied the hobo, "Neither have I."

As a matter of fact the hobo was merely trying to be polite when he professed ignorance of the meaning of the terms he had used. He knew perfectly well that "gazebo" or "gazabo" was a term which might refer to any member of the great fellowship of woods workers; that the "pike" was the supply road which led into camp; that the "push" was the camp foreman; that "sky-hooking" or "top-loading" was the work of stacking and arranging the logs properly on top of the car as they were loaded; that the "ground-hog" was one of the men who directed the course of the logs with his peavy as they were rolled up on the car; that a "blue" or a "blue-butt" was a twelve-foot log larger at one end than at the other and thus prone to roll up faster on the large end than on the small; that to "Saginaw" a log was one way to retard the large or butt end and to "Saint Croix" a log was another way to help the small or top end gain; that "gumming" a log was failing to keep the two ends even and having the top or small end go up first; and that getting his "stem" cracked was breaking his leg.

The hobo lumberjack probably also knew more of the picturesque slang of the camps, where dinner in the woods was called the "flaggin's," where bread was "sun-toast" and butter was "salve"; where the bills in the pay envelope were either "long greens" or "hay"; where hats were "sky-pieces" and where, if a lumberjack was about to die and wanted a priest or minister, a "sky pilot" was called to "give him a farewell start for Heaven."

But the average reader will be much in the same position as was the sister when it comes to the complex and diverse terminology which has sprung up in the logging industry. Obstacles as different in nature as the localities which breed them have led to the coining of a multitude of terms which are absolutely meaningless to the ordinary person. The methods of logging have been constantly changing and improving and these changes and improvements have continually given rise to new and puzzling terms. In an effort to make myself as clear as possible I will refrain as much as I can from the use of terms which would be obscure to the average reader in telling of the logging methods with which I have been familiar, methods which are now somewhat outworn in some localities but still obtain in others.

Let us view in brief entirely the course of the pine tree from cone to consumer and in so doing we will gain a fair understanding of the methods by which this country has been denuded of vast stretches of virgin pine timber within the last few decades.

Many ripe cones drop from the mature trees but relatively few of them reproduce. One falls where a shaft of sunlight beats down upon it, caressing and warming it until its scales open and allow the imprisoned seeds to escape. The seed which we shall follow works its way through the carpet of dead needles to the moist, warm soil beneath. Motivated by the universal urge to expand and grow and encouraged by the moisture around it and the warmth from above, it strikes roots into the soil and rears a head above the soft, woodland carpet. Through the long winters which follow the little pine is dormant, but in the summers it grows faster and faster. Finally it holds its head above the deep drifts of snow which have hitherto covered it in winter. In summer its roots strike ever deeper and farther afield in search of food and water, while its head yearns toward the sunlight and air of the upper regions, the open spaces above the leafy ceiling. In consequence of this yearning, its body or trunk grows straight and tall and strong with but few knots to mar its symmetrical surface.

Finally, after many years of constant, untiring growth, it lifts its head to the majestic level of maturity and looks down in tolerant contempt upon the struggling infants below. It has become a lofty, full grown pine, one of the most glorious creations of that Thing or Law or First Cause which we call Nature or God. It has achieved Nirvana. All its days are bound up with serenity. Serenely it absorbs the thirst-quenching, life-giving rains of spring. Serenely it weathers the hot suns and violent storms of summer. Serenely it gazes in autumn with pitying superiority upon its lesser neighbors which must shed their leaves and go garmentless and gaunt through the cold and seemingly interminable days and nights of winter. Serenely it faces and triumphs over the harsh winds and snows of that trying period.

And then, one day, the great pine senses the approach of tragedy, the beginning of the end of its serenity. A little company of men comes into the depths of the forest. Men it has seen before. Indians slipping on noiseless feet through the forest aisles. French *coureurs de bois* and *voyageurs,*

singing their rollicking boat songs as they make their ways
along the lakes and streams of the wilderness—trappers,
hunters, cruisers. All rather quiet men, using the wilderness and not seeking to conquer or destroy it. But these new
men are different. They are far from quiet. They shout
out healthy, hearty oaths and it would seem from their talk
that they intend to cut down all these trees. There is much
talk of a so many million cut this season. The lordly pine
sniffs in contempt.

These men take themselves and their work seriously.
They select a site for a camp, clear the ground, and construct
several buildings, a men's camp, a cook's camp, a stable, a
hay shed, and a granary. While they are engaged in this,
their leader, the foreman, has been prowling about the forest
making blazes on certain trees with his axe, laying out roads.
A main road is laid out along a creek bottom if the nature of
the terrain permits, so that there will be ample water at hand
with which to ice it. And the land along the creek is usually
level and devoid of obstructions and troublesome grades.
This main road varies in length with the size of the operations, sometimes being six miles or more long, and is laid
out with the greatest care. From it branch roads stretch
out like the limbs of a tree, covering the land to be logged
with a network of thoroughfares, all of which lead into the
main camp.

When the foreman has finished his blazing and the camp
buildings are completed, the axemen begin the road cutting.
They cut all trees in the roadway, which has a width of about
four rods, down to the surface of the ground. After completing the main road, they turn their attention to the tributary roads. Certain places along the roads have been picked

out as skidways and at these places a small clearing is made in which two logs are placed parallel to each other and at right angles to the road. On these timbers the logs are piled after they are cut and before they are carried to the landing. Horses haul the timber from the roadway, the foreman stakes out the road to a width of eight feet on the straightaway and more on the curves, and the scene is prepared for the entry of the grading crew.

The grading crew is made up of men unskilled in the handling of axes, but fairly adept with hoes, picks, and shovels. They follow the foreman's stakes and level the road to the best of their ability, chopping out roots, shoveling away humps, throwing out rocks and blasting boulders. In the rapid and efficient construction of roads no body of men on earth can excel a skilled logging crew.

Our heroic pine, all this while, has been growing nervous. The assurance and confidence of these men is unconquerable and the great pine begins to fear for itself. It has seen many great trees fall before the relentless march of these men as they built their roads and now it sees the beginning of a wholesale slaughter.

A logging camp is a flexible organization. The division of labor is not too closely defined and the lumberjacks take upon themselves various duties as opportunity offers and efficiency demands. In general, however, the arrangement which prevails is somewhat as follows. Five men and a two-horse team usually constitute a unit for the actual manufacture of timber from trees. There are several such units in every crew, the number, of course, depending on the size of the crew. It is the duty of two of these men, our pine tremblingly notes, to fell the timber. The great tree's first

sniff of contempt at the thought that such insignificant crea-
tures could cut down the proud product of many years of
Nature's handiwork has long since given way to fear and
belief. It has heard the descending crash of their victims in
the distance and now it sees their destructive operation at
first hand. They have with the axes, shining pieces of keen,
hard steel on long, wooden handles. And saws, long blades
full of many sharp teeth, with a handle on each end. These
biting tools go into play and forest monarchs topple about
our doomed pine, filling the air with their swan songs as
they crash quiveringly to earth.

Finally, the two woodsmen reach our pine. They quickly
decide which way it shall fall, taking into consideration the
direction in which the tree leans, if it does lean, and the best
open space in which to bring it to earth. On that side toward
which they want it to fall they take their stand, facing each
other, and their keen axes bite with steady regularity, one
after the other, into the tree until a sizable cut has been made
in its tough, dark hide and the white wood beneath. Then
they take their shining crosscut saw and attack the other side
of the tree, drawing its sharp teeth across the trunk at a
point an inch or two or three or four above the lowest point
of the axe cut on the opposite side. The teeth of the saw
bite insistently into the soft wood and the pine begins to understand why it and its fellows have always soughed sorrow-
fully in the winds of all seasons with an undertone of appre-
hensive pain. It must have been in instinctive anticipation
of this. In a final, desperate effort at self-preservation it
leans down heavily and tries to stop this hellish instrument of
torture which is tearing its vitals. But the effort is useless;
the stop only momentary. A heavy, steel wedge is inserted

in the cut back of the saw, which has now penetrated more than its width into the tree trunk. Strokes from a heavy sledge drive it into place and force open the cut. Oil is sprinkled on the saw blade and the work goes on.

As the teeth of the saw reach a point almost above the inner limit of the axe cut, there is a premonitory crack. The sawyers look up at the great pine, swaying a bit as it musters all its forces in a final effort to remain standing. Again its effort is in vain. A few more rapid, relentless strokes of the saw and there is another louder crack. The cry "Timber!" echoes through the woods and everyone within the danger zone scampers for safety, the sawyers with them, carrying their saw with them or leaving it in the cut, as wisdom demands. With a swishing crescendo culminating in a crash, the great pine falls to earth, perhaps tasting a sweet revenge as it pins one of its destroyers beneath it and hears his cries of mortal agony.

The monarch of the forest now lies prone and helpless and the work of dissection begins. The bright, sharp axe blades lop off the upper limbs of the tree and the saw divides the great length, sometimes one hundred and fifty feet, into a number of logs of specified size. These are dragged, one at a time, along "travoy trails" made by the "swamper," on a "travoy sled" pulled by a horse, or horse team, to the skidway. The teamster is assisted in his work by a helper known as the "chain man" or "chainer." One end of the log drags on the ground. The other is rolled onto the "travoy sled" with a peavy and chained in place.

Arrived at the skidway, the logs are piled up, tier upon tier, being rolled to the top of the pile by a block and chain arrangement powered by horses. There they rest until a

logging sled comes along. This is a simple affair consisting
of two sets of runners with a "bunk" on each. The first log
loaded is left in the middle to balance the "bunks." Then a
log is chained to each side, after which logs are packed between and the loading begins. The sled is loaded as high and
heavy as the condition of the road and the drawing power of
the horses will stand. When completed, a wrapping chain
is bound tightly about the load to keep it from spreading and
it is pulled along the icy roads to the landing or banking
place. There the logs are put in great piles on the river-
bank to await the coming of spring when the blocks which
hold them in place are pulled out, and they are tumbled into
the river and run downstream to the mill booms. There they
lie in waiting until claimed by the insatiable saws which cut
them into many shapes and sizes of lumber. The lumber is
shipped to all parts of the world, it is used for every con-
ceivable purpose, gives service for many years, and finally
disintegrates and decays and returns to the earth from which
it came. So ends and begins anew the epic story of the pine
tree.

PART II

WOODS-BOSS, CRUISER, AND CONTRACTOR

Chapter VI Jams and Washouts, Whiskey and Indians

I N the fall of 1877 I was employed by C. T. Pendleton and Henry Sargent as woods foreman. I was then only twenty-five years old—rather young for a foreman—but I had had sufficient experience and was tall and heavy enough to inspire a healthy respect among the men. We went into the woods about the first of October and I was busy there and on the river until the following July, a nine months stretch of the hardest sort of work and worry imaginable. This was on a tributary of the Oconto, named by the Indians the Waupee. The following winter we worked in the same territory.

That winter of 1878-79 there was an unusual scarcity of snow and as a result the spring was very dry, which prolonged the log drive and was very hard on the men. There was a great log jam that spring at the junction of the north branch of the Oconto and the main river. Several million feet were tied up, and the milling companies were anxious to have the logs brought through so that they would not have to shut down their mills. One head of water was left in the river above the jam, and that was in the reservoir dam known as the Shute Dam.

73

It was a ticklish situation and the contractors came to me and asked my advice as to the best way to handle it. I told them I could take twenty-five men and, with the water in the Shute Dam, run the logs to the Lindquist Dam about forty miles downstream from the jam. "Go to it!" they said and go to it we did. We broke up the jam a bit and then let the water loose and soon had all the logs in motion. For two days and nights my crew and myself worked tirelessly, with no sleep and with little to eat, there being no time for either. I always kept two of the rivermen together, so that if one was unlucky enough to fall asleep and fall into the river the other could fish him out. It was unlikely that two would succumb to Morpheus at once.

Perhaps it was on this drive, although I believe it was several years later, that a rather laughable incident occurred in which the joke was on me. I always drove my men pretty hard, for I knew that men had more respect and worked better for a boss who was not easy going. But it seems that, as is usually the case, I had a reputation for being much harder than I really was. One of my rivermen fell in the stream once and was under just about long enough to repeat the Lord's Prayer—which he probably did. When he was pulled out and restored to an interest in life, the first thing he gasped out to the man who had saved him was: "For God's sake, don't say anything to anybody about how long I was absent from work! If Nelligan heard about it, he'd dock me sure for the time I was under water!"

At the end of two strenuous days we had all the logs floating behind the Lindquist Dam, the water in which, when the gates were opened, carried the drive down past Oconto Falls and to the Stiles Dam, ten miles from Oconto, and then on to the booms. The greater part of the job done, we caught up on sleep and grub and returned to the upper works. Fortunately, there were heavy rains then and we experienced little difficulty in getting the balance of the season's cut down river.

Many amusing as well as tragic incidents would happen during the course of a log drive. When the water was not too swift and the drive moving along without trouble the rivermen liked to hold three or four logs together with their peavies and float along on this make-shift raft, as it was less dangerous and troublesome than riding a single log. I saw a bunch of them that spring sailing serenely along in this manner when they were suddenly confronted by a large, leaning maple tree, the branches of which brushed the water. There was no time to steer the raft around it, so it was a question of climb or swim and the lumberjacks indicated their simian ancestry by climbing. They clung ungracefully to the branches of the tree and shouted for a boat, which finally came along and picked them off, the strangest and most motley harvest of fruit a maple tree ever bore. It was an immensely funny sight to see those six lumberjacks clinging to that leaning maple tree. For a lumberjack is as strange to tree climbing as a sailor is to swimming, and looks equally incongruous when engaged in it.

There was in my crew that year a young German named August Schwartz who came from a farm somewhere in Wisconsin and had worked in the woods for Anson Eldred the previous winter. When spring came he decided to try river driving. Rivermen received two and one half dollars and four meals per day and this looked much more attractive to young Schwartz than working on a farm for sixteen or eighteen dollars per month. So he hired out in my crew and went to work. He was a husky young fellow and a willing and steady worker so he soon won our respect. Someone nicknamed him "Paddy" and the name stuck, probably because he was so unlike an Irishman. He had never had any experience at river driving and all the members of the crew were helping him to learn the dangerous game.

We finally arrived at Oconto Falls, after running the logs through the Lindquist Dam and there we began cleaning up the timbers which were stranded around the head of the falls. Paddy kept working closer and closer to the danger spot, where a tremendous volume of water thundered over the edge and took an abrupt plunge of forty feet to the riverbed below. I warned him repeatedly of the danger there and told him to keep away from the place and let more experienced men do the dangerous work. But he persisted, probably feeling that he should share the danger with the rest of the crew. The inevitable finally happened. Paddy made a misstep, was thrown into the terrific current, and carried over the falls before anyone could raise a hand to help him.

We were all quite dumbfounded, stood paralyzed for a time. When we regained our wits, we realized that it was useless to have any hopes. No man, we were sure, could live after going over the falls and being battered about in the seething caldron below. He would be either lost under the wing dams, or smashed to bits among the rocks. We all felt the loss of Paddy keenly, but the work had to go on and we continued, silently, thoughtfully, and perhaps a bit more carefully.

The buildings and camp area of a typical logging camp of the 1920's. The living quarters are on the right, with the cook shack and sawyer's shack on the left.

The rivermen taking the chance to eat a meal at the wanigan during the spring river drive. The wanigan housed the cook and all of his supplies and needs. This shack on a raft was floated along behind the drive to make sure that the men had at least two hot meals a day.

These two pictures above and below show the camp cooks and bakers at work in their kitchen area. On the cooling board you can see the mountains of cookies and loaves of bread that took to feed the hungry jacks after a day in the woods on the working end of a long cross cut saw

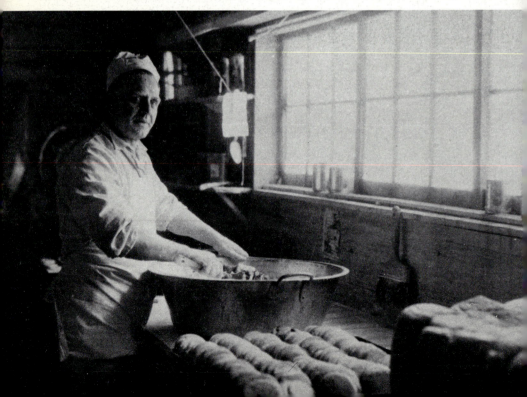

awyer at work putting finishing touches on saw. Note the rather rude, but effective sawyer's ench at which he is working.

A teamster is skidding a log from cutting area to stacking yard, using a snow path to ease the work of pulling the log over the ground.

Swampers and skidders rolling a log up the skids to the top of the pile. The top man will work the log straight with his cant hook as the teams of horses pull the logs to the top of the inclined skid. This picture was taken in a snow storm during extremely cold weather.

About an hour later Paddy appeared on the scene again. We stared at him in awe, for it was like welcoming a person back from the dead. His clothes were torn to shreds, but his bones were unbroken and, aside from the shaking up he had received, he seemed none the worse for his experience. It was little short of miraculous. He had been swept over a forty foot falls into an inferno of water, had been carried down a treacherous rock-studded rapids a mile long, and still had come out alive and unbroken. His tremendous vitality and strength, the high water, and plain, blind luck had combined to save him.

"I'm all right, boys," he said, in a voice which sounded a bit shaky, "but I lost my hat."

Mr. A. Cole, superintendent of the Holt and Balcom Lumber Company happened to be there.

"Paddy," he said, "when I get back to Oconto, Ill find you the finest hat money can buy and send it up to you!" And he did. Paddy wore it with great pride.

In the fall of 1879 I entered the employ of Sargent and Pendleton, a logging partnership of Oconto, as a camp foreman. The firm operated that season along the south branch of the Beaver River, a tributary of the Peshtigo. Four camps were opened and started operations along the stream on the first of October, I having charge of the one farthest upstream, close to the first lake. This little body of water was named for me, Nelligan Lake, and still bears that cognomen. It was spring and rain fed and had no inlet. A small creek served as outlet.

Across this small stream we threw a dam and, when it was completed, dropped the sluice gates and kept them down all winter, from November to April. We had hoped to get a good head of water back of the dam but were disappointed, the water rising only four feet during those five long winter months. My crew made a cut of three million board feet of timber during the season, all of which was landed on the lake.

When spring came we found that there was not sufficient water behind the first dam to carry the logs down to the second one. So we built two more dams in between and then cut a ditch between Nelligan Lake and another small lake in close proximity, draining the water from the latter into the former. This gave us enough water to start the drive, which got along fairly well for some distance downstream and then hung up high and dry for lack of water. This necessitated the building of another dam. When it was completed and all of the dams were filled to the highest point possible, the company prepared to complete the drive. Men were stationed at all the dams to sluice the logs and at bad spots along the stream to prevent jams and keep the timber running smoothly. Everything was in perfect readiness.

But about that time Nature, and Fortune, took a hand in the matter. The dams had all been built on a quicksand bottom and building dams on such a foundation is never very safe, no matter what pains may be taken in their construction. Even if piling is driven all around the foundation, it isn't safe. Like the owner of a car with worn out tires, you never know when you'll have a blowout. The reservoir dam, behind which a great head of water had accumulated, gave way first. The big bulk of water which it had held rushed downstream and threw itself against the second dam, which was unable to stand the strain and also gave way. The now greatly augmented flood continued on its mad course and carried away the third dam without any trouble at all. Increased still more in size, it rushed down upon the fourth dam.

I was at work with a crew of men repairing booms behind the fourth dam. I sensed or heard something wrong and looked up the river to see a mountain of water tearing down upon us, carrying uprooted trees and sawlogs on its crest as if they were matches.

"Get to hell off the booms!" I shouted at the men. "Run for high land! Hell's broke loose and the devil's changed the fire into water!"

They were quick-thinking, quick-acting rivermen and it didn't take them long to see the danger and get themselves out of the way. The booms were clear within a few seconds and we were standing in safe places watching the flood. It tore through the fourth dam as if there'd been nothing there, but by that time there was so much dam wreckage and so many trees and saw logs riding the crest of the flood that everything got mixed up in a mad mess just below the site of the fourth dam and jammed tight.

There the logs lay until the following spring. The company had to rebuild all the dams and it cost them a good deal of money before they finally got their logs to the mill booms. A bad mistake was made in not hauling the logs down-stream from my camp before banking them. A fine four mile road could have been built and the logs hauled by horse teams. Things were ideal for the construction and use of such a road. There was a grade of about three percent along the stream and there would have been lots of water with which to ice the surface. Had that been done and the logs put in the river a few miles below where they were, there would have been no trouble. The extra dams would not have had to been built and the drive would have gone through at the first attempt.

In the year 1882 I was sent out by C. T. Pendleton to cruise timberland in the vicinity of Monico Junction. The junction was then the construction headquarters for the line of the Chicago and Northwestern Railroad which was being pushed northwest from there through Rhinelander and on to Ashland. It was a canvas village and typical of the towns that spring up along a railroad in course of construction. The building of the road had progressed but a short way beyond there and it was still the end of the line in so far as regular train schedules were concerned.

I had with me two assistants. We had been out some weeks and our work was nearing completion when our grub supply ran dangerously low. We saw that we would be unable to finish our work before we ran out of grub, so I gave one of the men, named Harry, sufficient funds with which to purchase ham, flour, sugar, tea, coffee, condensed milk, and baking powder and sent him to Monico Junction with directions to buy the supplies and hurry back, for I had no intention of going hungry in the woods if it could be avoided. We were constantly moving in the course of our cruising, so I told Harry the section corner at which we would be encamped on his return and he started for town.

The other fellow and myself continued our work and arrived at the designated meeting place. We ate up the rest of our grub and Harry had not yet arrived. Two food-less days passed, we were still hungry as the wolves that we heard howling in the woods, and still no Harry. On the morning of the third day the belated messenger appeared on the scene without any food. He was pleasantly intoxicated, had a partially filled whiskey bottle, but not a crumb of anything like grub. Temptation had proven to be too much for him after a long period without liquor and when he arrived at Monico Junction he had promptly gone into a saloon and spent our grub money for whiskey. I leave to your imagination the warm, blue reception he was given. Experience is a harsh but good teacher and Harry, no doubt, learned a few new cuss words that day.

We packed our tent and camping outfit and started for the railroad right of way on which there was a boarding camp for the railroad builders. At noon we reached the camp and there ate a hearty dinner. Then we hiked on to Monico Junction, arriving there that evening all pretty well tired out. Both the other man and myself had enough money to pay our fares from Monico Junction to Oconto, but Harry had left not only the grub money but his own in the saloons and he was penniless. He asked what he was to do and I told him very plainly that he could go to hell for all of me and that if he wanted to reach Oconto, he'd have to walk. When we arrived at Oconto, I met Mr. Pendleton. He asked where Harry was and I told him what had happened and how to reach Harry by mail. He sent him money to come home on.

Forty-four years later, in November, 1926, I had to wait six hours at Monico Junction to get a train to Watersmeet. I found that it was not as lively a place as it had been back in 1882. The day was cold and there was no fire in the depot. I couldn't even find a drink of water during the six hours I had to wait. I trust that in forty-four years more I shall be in a warmer place.

Harry's little escapade reminds one of the classic joke that is told about the logging crew that ran short of grub in the middle of the winter. Several jacks were selected and given the necessary money to go to the nearest town and purchase sufficient supplies to carry the camp through the rest of the season. They went but, like Harry, their tastes ran more to liquor than food. When they returned to camp they had several cases of whiskey with them, but the only thing they had in the line of grub was a few loaves of the staff of life. Whereat one of the lumberjacks wittily inquired:

"What're we gonna do with all that bread?"

As cheap and easily obtainable as it was in those pre-Volsteadian days, great importance and value was attached to whiskey. There was the typical case of the Irish farmer who lost his home by fire. Everything he had—house, barns, animals, farm machinery, and everything else—was consumed or ruined by the flames, and the family was left destitute. Being rather pious by nature, they knelt in prayer, devoutly thanking God that the catastrophe which had befallen them was not greater and asking for divine help in the task of reconstruction. The head of the family calmly looked over the ruins of his life work, spent a few moments in prayer, and then turned to his wife and exclaimed with great fervor:

"Thank God, I've got two quarts left!"

The insatiable thirst that the lumberjacks and woodsmen of a few years ago had for raw liquor sometimes resulted in tragically humorous mishaps. A decade or so ago, Bob

Starr, with whom I was acquainted, and two other men were getting out tie logs in St. Louis County, near Duluth, Minnesota, on a piecework basis. They felt the need of stimulant, so they wrote a message to a druggist asking him to give the bearer "one gallon of good alcohol" and sent it in to the closest drug store by an Indian. Whichever of them wrote the note was a dreadfully poor writer and the druggist interpreted the message as calling for "one gallon of wood alcohol." So he sent a gallon of the deadly liquid back with the messenger. The tieloggers took no time to sample or test it but, secure in the belief that it was good alcohol, immediately started drinking. They were all poisoned and found dead sometime later.

Chapter VII   Cruising in Minnesota

O N the first of May, 1883, William Morrison, an experienced timber cruiser and myself left Oconto for Minnesota to look over and estimate timber lands in the region around the headwaters of the Mississippi. We had an agreement with Mr. James F. Connif by which he was to furnish our supplies, maps, and equipment and enter the timberlands we looked, and we were to have a quarter interest in the property. We went by rail to Brainerd, Minnesota, and there hired a team to haul us to Leech Lake, which lies in the Indian reservation, where we made our headquarters camp.

The country thereabouts is covered with lakes and streams and we did most of our traveling in a birchbark canoe which had been supplied us by Mr. Connif. We hired an Indian from the reservation to run the canoe, make trips to the headquarters camp for supplies, and do the camp work —with the exception of the cooking, which I did myself. The Indian turned out to be one of those sad specimens which are usually and unfairly considered to be typical of the race. He was an utterly worthless whelp. He wouldn't do a bit of work and was continually wasting his time loafing with other Indians.

On the fifth day out we picked a campsite and, telling the Indian to pitch the tent and cut wood, we went about our work of locating and estimating. When I returned to camp later in the day, I found that he had done neither of the chores assigned to him. I gave him a good tongue drubbing and be began to get sassy. So I lost what little patience I had left and told him to get to hell out of there, emphasizing my command with a few of the swiftest and best placed kicks which ever propelled an Indian on his way. He departed in anger and with many threats to the effect that he would get even with me one day.

About a month later I walked into Brainerd, a hike of about eighty miles, with our reports on the timber we had located. When I was a short distance from an Indian village, I saw my erstwhile Chippewa employee approaching. Indians have a habit of remembering things and I thought that the fellow might remember his threats as well as I did. I prepared for action, for I felt that he might welcome this opportunity to have revenge when he was so close to a number of his own people. It didn't take me long to decide what to do and how to do it. That was to knock him senseless just as hard and fast as I could if he showed any signs of fight. If he didn't wake up, it would make little difference. In those days no one ever thought of an inquest being held over a dead Indian.

As we neared each other, he showed no particular signs of recognition, nor of memory of the indignity he had suffered at my hands. I saw that he was in that state of drunkenness in which all the world takes on a rosy hue to an Indian. A red man can be very happy when he has just the right amount of firewater in him. But it must be neither too little nor too much. As he passed me, he looked at me through bleary eyes and murmured in the best of good humor, "Beshu-beshu!" which is the Indian equivalent of "Good day!" "Beshu-beshu!" I answered and continued on my way, thankful that the affair had terminated as pleasantly as it had. I went on to Brainerd and there mailed my reports to Mr. Connif, who was in St. Cloud, Minnesota. After a day of rest in Brainerd, I returned to Leech Lake where I was met by Mr. Morrison with the canoe and supplies. We then started on another two weeks cruise.

During the six months we spent in that region, I had much opportunity to observe the Indians and learn their racial characteristics. There was an Episcopal church on the reservation and I got rather well acquainted with the minister, a Mr. Wright, who has been a missionary among the Indians for forty years. He told me that he had little trouble with the thoroughbred Indians, but those who had a mixture of white and Indian blood were rather hard to keep under control. He said that shocking conditions followed when the white men got drunk and mixed with the squaws. Social disease throve upon the unbridled license which followed and spread to such an extent that the dogs became diseased from licking sores on the limbs of the Indians.

On the other hand, miscegenation seemed to be a success where the Negroes and the Indians mixed. In the years before the Civil War, I was told, the rich people from the South were accustomed to make journeys to the upper Mississippi country, bringing their slaves with them. When the negroes got in the northern states, where slavery was against the law, many of them deserted their masters and settled there. Some of them mated with the Indians and the half breed results of such unions were called "Bungoes." These Bungoes, as a general rule, seemed much brighter and more intelligent than the pure bred Indians. I met several of them at Leech Lake where they had stores and hotels.

Most of the Indians could understand English and talk it, if they wanted to, but they seldom wanted to. They preferred instead to pretend entire ignorance of our language and talk by sign language. If you asked them the distance to a particular spot and it chanced to be five and one half miles away, they would make marks on their five fingers and cross another one in the center.

They loved to gamble and poker seemed to be their favorite game. I watched six of them play poker once for six hours without an instant's let up. They played for money, but when any one of them ran out of filthy lucre, it didn't stop his playing. His possessions went into the pot then, his ponies, his clothes, anything he owned. A mackinaw which had been staked changed hands several times while I watched. They were very stolid and unemotional and their losses or gains made absolutely no impression on their stoic countenances. They had perfect "poker faces."

They were inclined to be both thieving and superstitious and I remember hearing of an amusing incident which happened only a few years ago near Duluth which illustrates these characteristics. A white trapper had his camping place situated close to a much traveled road in St. Louis County. He was encamped in a tent and had an outside fireplace. The Indians who passed on the trail would take possession of the place during his absence on the trap line and use his fire place and burn up all the wood he had prepared. Finally the trapper grew sick and tired of cutting wood for the use of uninvited guests. He buried a stick of dynamite beneath the place where the fire was usually built. A cap and fuse was attached and the fuse was left projecting a little above the ground, but was covered with inflammable rubbish so it couldn't easily be seen. Having set this little human trap with all the cunning of his kind, the trapper started out on his regular trap line. The Indians arrived and immediately built a fire out of the trapper's wood and made preparations for their meal, placing several kettles over the blaze. The fire was hardly started, however, when the fuse was ignited and exploded the stick of dynamite. The fire and kettles were blown to the seven winds by the blast and only a big hole remained in the earth. The Indians were frightened out of their wits. This was "bad medicine," indeed. They left that place in haste, and the trapper was never again troubled by them.

An incident occurred in 1883 which illustrates the devious methods employed by lumbermen in early days. This particular effort was eventually thwarted, but there were many such which were successful. The Minneapolis lumber barons needed water and lots of it to float their logs to the Minneapolis mills in the spring time. They were not, of course, allowed to build dams on the Indian reservation for such a purpose, but they succeeded in putting a bill through Congress providing for the erection of a dam on Swan River, the beginning of the Mississippi and the outlet of the three lakes, Cass, Leech, and Winnibigoshish, on the ground that the water stored up by such a dam during the winter months would help navigation on the lower Mississippi during the summer. The power which the lumber barons wielded in Congress is evident when one considers the flimsy pretext on which they put this measure through. The idea of even the largest possible head of water up there helping navigation on a stream as large as the lower Mississippi is nothing less than ridiculous. But they passed the bill and began construction of the dam. The Chippewa called a council of war in October, 1883, and asked the Sioux to join them. About one hundred of the plains Indians attended. A war dance was started and kept up continually for three days and nights. Then all the dogs were killed and their meat thrown into a grand stew which was eaten by the braves with spoons and tin cups. The squaws, in accord with long established tradition, stood aside and took no part in the dancing or eating of dog meat. Finally the council came to a climax and about five hundred Indians marched to the place where the dam was under construction and peremptorily called a halt to the building, rightfully claiming that the backing up of the water would ruin the wild rice and the fishing by stagnating the water. And that was as far as the dam got.

We worked steadily from the first of May to the first of November and located and estimated about seventy-five million feet of the very best pine timber in Minnesota. When our work was completed, we returned to Oconto. Mr. Connif suffered financial reverses and couldn't raise the necessary money to enter the land. So we never received a cent for our summer's work.

Chapter VIII   The Harmon Company

SHORTLY after my return from Minnesota, I made a trip to Canada to visit my mother, whom I had not seen for a number of years and who was growing so old and restless that it was difficult to keep her contented in one place for any length of time. An amusing thing happened as we neared the Canadian line. The customs officer came through the train coaches and ordered all the passengers to prepare their luggage and belongings for inspection. I was opening my grip when I noticed that the lady who occupied the seat directly back of mine, evidently a Canadian woman returning home from a shopping orgy in the States, had lifted up her voluminous skirts and put a roll of silk, on which she wished to escape paying duty, beneath them.

The skirts, made up in the style of the time, seemed quite ample enough to have easily concealed a baby elephant, so I made a facetious suggestion that it would be nice if she would be kind enough to hide my grip in the same place. But such levity did not appeal to the lady and she cut me off with a curt reply to the effect that I could take care of my own baggage. Her ruse was successful, for not even the most bold of the customs officers of that day would have dared to look under skirts for contraband goods. Now-a-days, whenever I hear of attempts to smuggle liquor across the border, I think of that lady and her voluminous skirts. What a blessing such garments would be to smugglers in this day and age! Two or three cases could be concealed under such skirts as those.

Immediately after returning to Oconto, I went into the woods for the Holt and Balcom Lumber Company, which had camps along the north branch of the Oconto River. I was employed by them all that winter. This company suffered a severe loss in December, 1884, the year after I worked for them, when four of their dams were washed out on McCauslin Brook, a tributary of the north branch of the Oconto River. The rains had been fairly heavy that fall and there was a twelve foot head of water stored up in the reservoir dam at the head of the brook for the spring drive. There were three dams below the reservoir dam on the brook and one on the north branch of the Oconto. Each of these was about half filled with water which had been left in them to freeze and provide good landing places for the logs during the winter, as well as sufficient flowage with which to handle them in the spring. Twelve feet of water proved to be too much for the reservoir dam and it went out with a roar. The flood swept down the riverbed, carrying everything before it, uprooting and snapping off trees, washing away the banks and raising a terrific, thunderous clamor. One after the other it washed out each of the four dams below it, added their contents to its own and proceeded on its mad march downstream, raising the water to an exceptionally high and dangerous level for miles along the lower Oconto River. The dams were owned by the Holt Balcom Company and the Oconto Lumber Company and represented a replacement value which was conservatively estimated to be fifty thousand. To build new dams would have been too expensive a project for the small amount of timber still to be taken out of that region, so railroad spurs were constructed from the closest main line and all the logs were railed to the mills.

In the spring of 1884, following the winter spent with the Holt and Balcom Lumber Company, I hired out to C. T. Pendleton and Son as boss of a crew of men on the east branch of the Sturgeon River, a tributary of the Menominee. I worked for them through the following winter and in the spring of 1885 I entered the employ of the A. M. Harmon Lumber Company as foreman. The Harmon Company was then operating on the east branch of the Sturgeon, and the White-Friant Lumber Company of Grand Rapids, Michigan, commenced operations on the same stream that spring. The Harmon Company had about ten million feet of logs in the river ahead of the White-Friant Company and the latter demanded that the river be opened, contending that it was a navigable stream. The Harmon firm then served an injunction to prevent the White-Friant Company from action. The injunction was sustained by the court and tied them up for that season.

The White-Friant Lumber Company was owned by Thomas Friant and T. Stewart White and was about the finest lumbering organization in that region. White and Friant were noted for having the best of lumbermen in their employ; for paying the best wages and supplying the best food and sleeping quarters; for the fine logging horses they had and the great pride they took in their work. The lumber they manufactured was railed to Escanaba, Michigan, on the Northwestern line and there loaded on boats which carried it down the lakes to the eastern markets.

When Flannigan and I commenced logging on the east branch of the Sturgeon, a few years after I left the Harmon

Company, White and Friant were still operating there.
During the spring drive our logs, which had to be driven
into the Menominee, got all mixed up with the White-Friant
company's timber, which was to be sawed at their mill at
Hardwood. They generously offered to help us out of the
mess, so I gave them full charge of our timber and our crew.
White and Friant had a river foreman named John Boyd,
one of the best logging generals in Michigan, and it didn't
take him long to thoroughly straighten out the situation.
He placed our men wherever he could use them to the best
advantage and soon had our logs separated from those of his
own company and our drive ahead of the White-Friant
drive. Everything worked beautifully. It was a typical
example of the splendid sportsmanship and coöperation of
the White-Friant Company and we appreciated it greatly.

T. Stewart White was the father of Stewart Edward
White, the noted author, whose works, *The Blazed Trail,*
and *The Riverman,* have done much to make immortal the
American lumberjack. When I was logging on the Pine,
Brulé, and Paint Rivers, all tributaries of the Menominee,
Stewart Edward White spent about forty days in the woods
with me. He hired out as a regular hand and did his work
capably and conscientiously. When I offered him his check
at the end of his stay with us, he smilingly refused it, saying
he had been working for his health. At the time I thought he
was a bit light-headed. I wasn't used to having men turn
down checks. But later I found out that he had been ab-
sorbing atmosphere and gathering material for the books
which were to make him famous. He was a fine-looking,
quiet lad and intimated to none of us that he was a writer,
cannily realizing that such a revelation would defeat his own

ends and render him a stranger among the jacks. Your American lumberjack is not an easy man to understand. But young White studied him and understood him and made him live in his books. He was very popular among the men, for he was a great story teller and such a man, if he is not too forward, is always well-liked by the men in a logging camp. In addition to a first-class school education he had, as is shown by his books, an extensive knowledge of practical lumbering, which he had gained, I suppose, from his father, who was a skillful and experienced lumberman.

My first job as foreman for the Harmon Company was to clear off the ground on which Foster City was to stand. This completed, we built some shacks for the men and cooks, constructed a dam, and began clearing ground for the mill and lumber yard. The Harmon Lumber Company's new superintendent, whom I shall call Gorman here, although that was not his name, arrived on the scene early in the summer. He was a life insurance man from Cleveland, Ohio, who held a small block of stock, about ten thousand dollars, in the Harmon Company, and his ignorance of logging and lumbering was immense and amazing. He couldn't tell a pine log from a hemlock log and, what was worse, his word wasn't worth, to put it politely, the explosive necessary to project it to hades. He got in bad from the first by building dams on land which belonged neither to him nor to the Harmon Company. The White-Friant Lumber Company then bought the land on which the dams were built and raised "merry hell" with Gorman. But in spite of such a boss, we worried along somehow and by fall Foster City had become a reality.

I made a short visit to Oconto and on my return trip to Foster City had a rather close call. I was sitting in the smoker when the train pulled in to Marinette and a number of passengers entered the coach. There were several young men bound for Hemlock River to repair some dams, one a big bum who was quite drunk and had reached the trouble-making stage. He began abusing one of the young fellows—a youth of only sixteen or seventeen years—and insisted that he could and was going to give him a licking. There were quite a few clean and sober woodsmen in the car and they finally interfered. But that only changed the direction of the bum's trouble-making efforts.

I was sitting rather close to him, talking to a man along side me and paying no atttention to the drunk, when suddenly, without the slightest warning, he jumped from his seat, pulled a large revolver from out of his pocket, cocked it, and covered me. His trigger finger looked a bit nervous and shaky, so I wasted no time but jumped up like a flash, caught his hand and held it in a safe position while I wrested the gun from it. I had fairly steady nerves back in those hectic days and had need of them often. Pulling a gun didn't greatly appeal to the other occupants of the coach and they were much inclined to kill the bum, but they finally let him go in peace, quite subdued. I gave the revolver to the conductor, Mike Eagan, and never heard whether the drunk recovered it.

We went into the woods for the winter and drove our winter's cut on the Sturgeon next spring. During the summer the mill at Foster City had to close down for lack of logs to cut. I had a crew of men on the river getting logs down to the mill and the work was very difficult due to the

low water in the stream.   The wild raspberries were luscious
and plentiful along the river banks and Gorman wanted me
to take part of my log driving crew off the river and put
them at work picking raspberries.   Think of it!   Here was
a lumberman, or at least one who claimed to be a lumberman,
with his mill closed down for lack of logs, who wanted to
take his driving crew off the river and indefinitely postpone
the re-opening of his mill for the sake of a few raspberries.
It was almost incredible.   The absurdity of such a proposal
should be greatly appreciated by anyone with the slightest
knowledge of logging and of the importance of taking down
the drive.   Of course, I paid no attention to the nit-wit and
went on with the logs.

We completed the drive about the last of October and
the driving crew, about seventy-five in number, was paid
off at Foster City.   There were no saloons there, so the first
thing the men did was to go to Metropolitan, where plenty
of beer and whiskey was available.   They stormed into Mike
Horigan's place and by the time darkness fell everyone was
pleasantly drunk and it was a merry party indeed.   In
order to liven things up a bit more, the man who had been
my cook on the river, a French-Canadian named Joseph
Gousaw or something of the sort, who was as good a fiddler
as he was a whiskey drinker, brought out his violin and
began to play, finding his audience very appreciative.
Everything was going along very sweetly when a fellow
named William Knox piped up and said:

"Let's hear 'The Protestant Boys', Joe!"

It was an unhappy suggestion.   "The Protestant Boys"
is a song of the Orangemen and there were several good
Irish Catholics in the crowd.   Hell broke loose automatically.

Two brothers named John and James Enright swore by all the powers of Heaven and earth that there would be no "Protestant Boys" played there that night and they commenced to clean the place out. A good many of those present including the poor fiddler didn't know what it was all about, but that didn't prevent them all from participating in a good fight. In a short time the place looked like a cross between a hospital and a morgue. The Enright boys were the victors for no strain of "The Protestant Boys" was heard on the evening air that night.

In the spring of 1889, three years after I left the Harmon Company, a laughable incident occurred at the dam which I had built at Foster City. The stream was so high that spring that the gates were unable to take care of the flow and the water began to flood the mill. To stop this Gorman had his foreman cut a gap in the wing of the dam. A laundry house, which handled the laundry in the settlement, had been built on that wing of the dam and was run by a man, his wife, and their half-wit son, who was sixteen or seventeen years of age. As the water rushed through the gap made by the foreman, it kept washing away the sides, and finally the laundry house with the imbecile youth in it was carried away on the flood. It was washed about a mile downstream and there lodged against the railroad bridge. A rope was thrown to the half-wit, who had been greatly enjoying himself the while by pounding gaily on a stovepipe and singing idiotic songs of his own composition. He was pulled to safety and then Gorman, with a characteristic lack of common sense, had his men set fire to the laundry house. As soon as the flames got under way the entire crew had to be set to work carrying water to keep the flames from burning the wooden railroad bridge.

Chapter IX Larry Flannigan

IN the fall of 1886 I left the employ of the Harmon Lumber Company and formed a partnership with Lawrence S. Flannigan, an alliance which was to last for many years. "Larry" Flannigan, as we called him, was the embodiment of all characteristics generally considered to be typically Irish. He was the most typical "wild Irishman" it has ever been my fortune to come in contact with, one of the most lovable and at the same time devilish rogues that ever walked the paths of God's green footstool. Aside from its business aspects our partnership was a continuous succession of pranks and practical jokes played by Flannigan on me, or vice versa, usually, I must admit, the former. In tribute to Larry and to the memory of our long friendship, I must here devote a few pages to some of our mutual experiences, incidents which, if thrown in elsewhere, might appear irrelevant.

Flannigan was the second cousin of General Philip Henry Sheridan, the dashing Union cavalryman who gained for himself a degree of immortality by his famous ride "from Winchester, twenty miles away." Larry's mother had been a first cousin of the General's. In common with "Little Phil," Flannigan had an unusual love for and ability with horses. He was a splendid rider and could handle the most unruly beast with perfect control. This was well-demonstrated the first spring we drove logs on the Fence River in the upper peninsula of Michigan. We had been "outside" and had received word that one of the dams had been washed out. Flannigan and I set out for the spot on two saddle horses. On reaching the dam we looked over the damage and put about twenty-five men and a number of horses at work repairing it. Flannigan's eye, always on the alert for handsome horse flesh, lit on one of the horses at the dam and he immediately decided to ride it back to the railroad station. Argument was useless when Larry had once made up his mind, so I let him have his way and he left his own horse at the dam and mounted his new steed.

We started back for the station. Flannigan's new mount was a splendid-looking, high-spirited beast which was inclined to be frisky and show signs of fight. But Larry, expert horseman that he was, had little difficulty in keeping him under control and we got along famously until about six miles from the station. Then Flannigan's love for a good joke overcame his usual good sense and, keeping his horse under tight rein, he worked him into a frenzy. We had been keeping some distance apart and he called out to me to pull up closer to him, saying that the Fence River Improvement Company would not stand the expense on the dam and that we would have to. Without the least suspicion I drove up close to him to talk the matter over. Then, without any warning, he suddenly jumped from his horse and turned it on mine—roaring like an Indian devil. I threw myself off just in time to escape death from one of the flying hoofs of Larry's horse, landed ungracefully in a puddle of mud, and dragged myself to a place of safety behind the trunk of a nearby tree. My horse set out for the station, hell-bent for election, with the other close on its heels. When they had disappeared, I looked around for Flannigan, but he had discretely faded from the picture.

If he had been within reach, I could have cheerfully killed him. I tried out my lame bones, found them still intact, and, stiff and sore, set out on foot for the station, where I arrived very tired in time for dinner. Flannigan followed me—at a safe distance. He turned up about an hour after I got in and located the manager of the boarding house.

"Is Nelligan here yet?" he asked.

"He is!" answered the manager, to whom I had told the story.

"Has he cooled down yet?" asked Flannigan.

"He has not!" replied the manager.

"For God's sake, get him away!" pleaded Larry, "I'm nearly starved."

The manager came in, told me about the conversation, and asked me to let him have his dinner without trouble. With my stomach full and the affair over without serious consequences, I began to see the humor of the situation, so I had him tell Flannigan that it was safe to come in and eat.

I was the victim of another of his practical jokes one year when we attended the Brown County Fair at Green Bay. The weather was fine and there was some great horse racing going on. As a result all the horse racing fans and sports in that part of the country were there. The town was crowded and we were forced to take a room at the Beaumont Hotel with only one bed in it. All tired out, I went to our room one night at about ten o'clock and turned in. Flannigan stayed out with a gay party of railroad and lumbermen—such parties are usually rather gay— and made his appearance in the room some time after midnight. Something was evidently resting heavily upon his conscience, for he knelt by the side of the bed and piously began to say his prayers before turning in. I was not accustomed to such conduct from Larry.

"Get into bed," I told him, "The Lord will not be paying attention to your prayers this night."

He was much insulted, being at that stage of intoxication where grievances are magnified.

"I'll get you for that!" he replied, in a grieved tone of voice, "and before tomorrow night."

I rolled over and went to sleep, with no further thought of his threat. About six o'clock the next morning he got up. I was still asleep so he fulfilled the threat. He set the bed clothes afire and calmly walked out of the room. I woke up with the covers blazing merrily all around me, jumped out of the fiery bed, grabbed my clothing, and ran into a salesman's room directly across the hall. The salesman got some water and put out the blaze. I dressed and went downstairs, saying nothing.

Shortly afterward the chambermaid found out about it, attracted to the room by the odor of the burnt bedding. Flannigan got hold of her and told her it was all my fault. He said that I was a bit crazy; that it would be well for her to watch out for me, as I had been running through the house without any clothing on; that he was going to have me taken before the court and have me committed to the insane asylum at Oshkosh. Then he thrust a five dollar bill into her hand and told her to hush the affair up. She grabbed the bill and shoved the burnt bedding down the laundry chute. I would have liked to report the affair to the proprietor, in which case Flannigan would have had to pay about twenty-five dollars, but I didn't like

to get the chambermaid in bad.   When the salesman asked
Larry what caused the fire he solemnly replied:

"It must have been a visitation of Divine Providence
on a wicked soul and nothing less!"

Our occasional stays together in hotels were usually productive of some such pranks as the one above related.   A
similar incident happened one Sunday morning in Escanaba
where we were staying together in a hotel room which, this
time had two beds in it.   Larry woke up about four o'clock
in the morning and wanted me to get up and go down to
Tommy Currie's place with him for a drink.

"Currie's won't be open this early on Sunday morning,"
I objected, not feeling like getting up that early on the
Sabbath.

"We'll take out one of the windows," countered Larry.

"I don't want to get up and I don't want to break into
Currie's and I don't want a drink and I do want to stay in
bed!" I replied and buried my head in the pillow.

Flannigan quickly jumped out of bed, took the large
pitcher of water which stood on the nearby washstand, and
pitched its entire contents all over me.   I was up in a flash
and after him.   Down the stairs and out into the street he
fled, with me close on his heels, both of us without clothing
except for short undershirts.   Fortunately for us, there was
no one on the street at that early hour.   I chased him all
around the block, throwing everything I could lay my hands
on at him but with little success.

Finally he stopped and pleaded:

"Let's make up or we'll be arrested!"

"I don't give a damn if we are arrested," I replied.   "In

jail I might get some protection from such a damn lunatic as you!"

But, as usual, we finally made up, went back to the room, dressed and got out. By that time Tommy Currie's place was open and we went down and had a few cherry ripes and laughed over the affair.

Once when I was returning home from the woods, I came in from Ellis Junction by way of the Chicago, Milwaukee and St. Paul line and was met at the Marinette station by Flannigan. He invited me to stay overnight with him as his wife was visiting in Fond du Lac, and I consented. Sometime during the night he got up, opened my handbag, took all my clothing out of it, and filled it up with some of his wife's intimate wearing apparel, "unmentionables," as they are aptly called. There was no necessity for me to open the bag and I left for Oconto the next day, entirely unaware of the trick which had been played on me. When I arrived home my wife took my grip, as usual, and opened it. She took one look at the contents and then confronted me with the opened grip.

"What does all this mean?" she demanded.

I stood speechless for a moment and then the origin of the trick dawned on me. I told my wife how it must have happened and asked her to wait until Mrs. Flannigan had returned from Fond du Lac and then phone her and ask her if she had missed any underwear. She knew Larry fairly well, so she did as I requested and, fortunately for my domestic felicity, Mrs. Flannigan was of a less prankish nature than her husband and told the truth.

Like most men of spontaneous deviltry, Flannigan was very generous and kind hearted. He mixed these admirable qualities with a shrewdness which made him a very keen business man and as good a partner as one could find. He put over a characteristic deal once in Marinette. Two men had been in business there as partners for some years, but they found each other's company less and less congenial due to religious differences. Finally they came to the decision that one of them must buy the other out. One of the partners was fairly well fixed financially and able to do this, but the other was without the necessary funds. The latter appealed to Flannigan and Larry agreed to back him and gave him instruction as to how to close the deal. The poor fellow went to his rich partner and asked him to place what he considered a fair valuation on the store. Feeling sure that he would be the purchaser, and not the seller, the rich fellow made the figure very low. Flannigan's man said he would think it over for a few days. He then got the necessary funds from Flannigan and bought out his partner, who couldn't kick because he had been making most of the trouble, had suggested that they sever partnership, and had been taken at his own word and figure.

All in all, Flannigan and I made a good working team and an uproarious brace of partners. Although he was usually the one who laughed, I sometimes turned the tables and played a good joke on him. The funniest of these was perpetrated when we were lumbering the Fence River in northern Michigan. We had been at different camps and had made an appointment to meet at the Forks Dam. On my way through the woods to the dam I met a saloon keeper from Marinette, engaged in collecting whiskey bills from the lumberjacks. He told me all about one of Flannigan's amorous indiscretions in that city. The story was good and

I turned it over in my mind and decided to make Larry suffer a bit. I reached the dam, fed and made ready the team, and was waiting for dinner when Flannigan arrived, tired and hungry. I told him the story immediately, included some excusable exaggerations, added some purely fictional dire consequences, and wound up by telling him that the sheriff was in the woods and that he was looking for him. He swallowed the story, hook, line, and sinker and, worried to death, set out for Witbeck on foot without waiting for dinner. I ate my dinner between fits of laughter, lay around for some time to give Larry a good start, and then kept behind him with the team all the way into Witbeck, a distance of about six miles. He soon found out that the whole affair was a hoax, but he took it without a word and never mentioned it until about a year afterward when he told me it was the worst, or the best, joke I had ever played on him.

Larry Flannigan was, if I may seriously use a hackneyed and much-abused phrase, "One of nature's noblemen." When he died at Beaver Dam some years ago, his son-in-law wired me and I attended the funeral and acted as pallbearer. As we wended our way through the cemetery carrying the casket, I said to the others:

"Well, Larry had the first joke on me and he got the last!"

How he would have chuckled over that epitaph!

Chapter X   Partnership Business

As I have stated in the preceding chapter, Larry Flannigan and I formed a partnership in the fall of 1886 and took a logging contract from T. A. Sheppard and Company of Chicago, which called for the delivery of two million feet of logs at Foster City the following spring. This timber was standing along the east branch of the Sturgeon River, a tributary of the Menominee. We gathered a small crew of men together, went into the woods, established our camp and got to work. Our initial venture as independent loggers was moderately successful. We completed the contract satisfactorily and had the logs delivered at the stipulated destination by the middle of May, 1887.

The Harmon Lumber Company was operating on the same stream and we drove their logs at the same time we drove ours. A few of the Harmon Company's logs were at the upper end of the stream and, due to the low water that season, we were unable to get them down and had to leave them there until the following spring. The Ford River Lumber Company also had some timber on the Sturgeon. They logged about five million feet the winter of 1887-88 and when they took down the drive the Harmon Company's leftover logs got all mixed up with theirs. I had been working on the Whitefish River and when I went up on the Sturgeon to clean up the leftovers, I found this to be the case.

The superintendent of the Ford River Lumber Company had the consummate nerve to put in a ridiculous claim against the Harmon Company for bringing their logs down

106

in his drive. I happened to meet him at the old, abandoned depot at Metropolitan. I called him in and shut the door.

"What do you mean by this bill you've presented?" I demanded.

His explanation wasn't very satisfactory.

"Cancel that bill in a hurry or I'll pound hell out of you!" I threatened him.

My bluff worked. He was scared stiff and didn't lose any time in canceling the preposterous charge. Mr. J. W. Fordney was with me at the time and he said that it was the most effective method of settling a claim he had ever seen. After that little encounter with the Ford River superintendent at Metropolitan, everything moved along smoothly and there was no trouble whatever.

Sometime during the winter of 1887 a man named Frank Burns of Oconto, broke his leg while working for the Mann Brothers on the Escanaba River. The company had two of his camp mates take him to Oconto for medical treatment. One of these men was Richard Doyle, who was one of our best foremen in later years, and was popularly known as "The Scout." Doyle and the other fellow brought Burns in on the train and then took him up to the doctor's office to be examined. Having completed the performance of his duty, Doyle promptly went out and got drunk. The doctor looked Burns over, and found him in such bad condition that it was necessary to amputate the limb. He did so and Burns died as a result of the operation. He was prepared for burial, but the amputated limb was left in the doctor's office by mistake and it was not until the final services were almost begun that the omission was discovered. Doyle, who was in attendance at the funeral, was

dispatched to the doctor's office to get the missing limb. He
hadn't wholly recovered from his drunk. He found the limb
all right, but he wasn't in quite the proper mental state to
consider the niceties of funeral etiquette, and he threw the
amputated leg over his shoulder without any covering over
it and carried it into the house as if he had a shank of deer
or mutton. Everybody was shocked beyond expression and
it just about ruined the funeral.

One day during the spring drive of 1887 I was standing
on the bank of the Sturgeon River near one of our dams
with James McGillan, Sr., of Appleton, Wisconsin. One of
my men had a fire near where we stood and was thawing out
some dynamite. He had two fifty pound boxes set up close
to the blaze, a little too close, as it happened, and the man
wasn't watching closely enough. It caught fire finally and
began to burn like grease. I noticed it, yelled "Run!" to
McGillan, and started for cover like a bat out of Hades.
We huddled behind a large pine stump some distance from
the place and waited for a tremendous explosion. But we
waited in vain. It was a lucky thing for us that there were
no dynamite caps near the blaze. They would have set off
the dynamite and everything in the immediate vicinity would
have been blown to bits. When we dared to look almost
all of the dynamite was burnt up. McGillan heaved a sigh
of relief.

"What would have happened if that stuff had exploded?"
he asked, wiping the sweat from his brow.

I grinned at him.

"There'd have been two strange Irishmen in hell for
first lunch!" I replied.

That summer, 1887, I went from Oconto to Bay City, Michigan, to look over some pine timberland on the Yellow Dog River in the upper peninsula. The timber was owned by the McGraw Brothers and they were planning to log it. From Negaunee, Michigan, I took the Duluth, South Shore and Atlantic Railroad to the Straits and Bay City. On our way we stopped at Seney while the locomotive took on coal and water. Seney was about the toughest town in the north country at that time. The lumberjacks were all on their annual spree and I could look out of the window and see about five hundred drunks on the main street. Two of them thought they'd play a little joke, so they boarded the train with revolvers in their hands and threatened everyone they saw. When they entered the ladies' coach the women and children almost died from fright. The men on the train could see it was only a bluff so nothing was done to stop them. The two drunks thought it was a pretty good joke, but the railroad officials didn't agree. They had the two practical jokesters arrested and the court gave them five years in Jackson prison to think their joke over and analyze its weak points.

That fall, 1888, when we were going into the woods, our crew got into an amusing bit of mischief. Flannigan and I left Oconto and Marinette the first of September with about seventy-five men. We went by way of the Chicago and Northwestern Railroad to Turin, Michigan, the closest station to our camps on the Whitefish. Almost every one of the lumberjacks had brought a bottle with him and by the time we reached Turin most of them were gloriously drunk. We got in there in the evening and were not to leave for the camps until early the following morning. As the lumberjacks began to sober up, their appetites grew keen and they looked around for something to eat.

An old Grand Army man named Brown and his wife had their home there and possessed a well-stocked chicken roost. The men found it, raided the roost and had the cooks prepare them a chicken dinner. Mr. Brown was away and when Mrs. Brown went out the next morning to feed her chickens, she found the flock sadly depleted. It didn't take long to determine the reason and then there was the devil to pay. Flannigan and I had gone to bed early and so knew nothing about it; the jacks had discreetly cleared out for the camps after their midnight meal. I finally told Mrs. Brown to count her chickens and put a price on the missing ones. She did as I requested and I paid the bill, a few dollars.

All of the same crew were in camp all winter and when I paid them off in the spring I deducted fifty cents from each man's check, a fair enough price for a country chicken dinner. I turned the thirty-seven dollars and fifty cents over to Mrs. Brown, who was delighted and would no doubt have liked to sell the rest of her flock at an equally high price. The men only laughed and didn't kick a bit when they found out what the deduction was for.

Our second winter under contract for Eddy-Glynn was very successful. We had two large camps, banked about eight million feet of timber, and completed the season's cut around the last of March. Flannigan and I each had charge of a camp and we vied with each other to see who could log the timber the cheapest. We each kept a very strict account of everything; men, horses, supplies, length of roads, daily cut, and so on. The timber was very much scattered in my territory and this made necessary a great deal of road cutting. I laid out most of the roads myself, blazing the trees plainly along the best routes so that the road cutters could make no mistakes. It was a strenuous winter. I often left the camp in the morning an hour before daylight and didn't return until an hour after dark.

We had five or six young Germans working for us in one of the camps on the Whitefish that season. One of them had a girl outside somewhere and was suffering from a bad case of love-sickness, which is the most dangerous disease. It finally got to the point where he thought he had to get outside and see her. He hated to admit the reason for his wanting to go, so he tried pulling a bluff on us. He fell in a mudhole and claimed that he had hurt his back. He said that he was in terrible pain and that he was sure he was going to die. We knew it was a bluff because he was short, stout, and husky and could never have hurt himself seriously by falling in the sort of mudhole he'd picked. But he wasn't much use in camp, and there was nothing to do but take him out. Flannigan hitched one of the teams to the lumber wagon, made a bed in the wagon for the patient, loaded him in, and was ready to start for the station. I threw a heavy pick handle in the wagon.

"Roll that German over and hit him in the back of the head with this pick handle before we reach the station!" I told Flannigan. "I'll go on ahead to Lathrop and wire the undertaker at Ishpeming to come down on the evening train with a casket, a black suit of clothes, a white shirt, and a black tie."

The poor German was scared out of seven years' growth. "Oh, God!" he groaned, "I ain't hurt that bad, boys!"

"You shut up," I told him. "We'll send you home to your folks clean and well-dressed. They won't think of looking at the back of your head and they won't have any expense in burying you."

He was speechless. I walked on ahead of the team and ordered supper for three. When Flannigan and the team arrived, it was time to eat and the patient got out of the back of the wagon and bolted as much as two or three ordinary men. We bought him a ticket to Green Bay, put him on the train and that was the last we saw or heard of him. The other fellows told us that he couldn't get the girl out of his mind and that he'd have surely hung himself if he had stayed in camp much longer. Whiskey is bad enough, Lord knows, but when a lumberjack falls in love, it beats whiskey all hollow.

That same winter, 1888-89, we contracted with J. W. Fordney of Saginaw, Michigan, to log, drive, and deliver to the booms at the mouth of the Ford River twelve million feet of pine timber. We had four camps on the Ford and its tributaries and got the timber out as specified in the contract. But when it came time for the drive, we encountered a little difficulty. The superintendent of the Ford River Lumber Company, the man whom I had threatened in the depot at Metropolitan, was not favorably impressed with my blunt methods of doing business and he would not have me on the river during the drive under any circumstances if it could possibly be avoided. He offered to drive our logs for less than one half of what we could do it for ourselves, so we readily consented. We were making easy money and none of our men were getting hurt. It must have cost him about two dollars per thousand to drive them, due to the dry season, and we paid him only seventy-five cents per thousand. He was probably bothered by the loss, for he tried to give Fordney a mean deal, but didn't get away with it. The Ford River Lumber Company's logs were in the stream ahead of Fordney's and the river was blocked by them for several miles above the mouth. Fordney, however, made them open a channel for him, ran his logs down into the bay, and there had them run into bag booms and towed to Marinette, where they were sawed into lumber.

J. W. Fordney was an all around good fellow and one of the cleverest timber estimators and salesmen in the entire country. As a boy he looked timber for Mr. Bowing of Detroit, and received for his work a part interest in all the pine timber he located on vacant government lands. He thus gained a very extensive knowledge of the timber lands in the upper and lower peninsula and the possibilities of marketing the timber on them. After we lumbered for him, he was elected to Congress and represented his district continuously for twenty years, becoming prominent as chairman of the Ways and Means Committee and joint author of the Fordney-McCumber tariff bill.

One of our camps that year was on the main Ford River and had about thirty men in it. The cook in that camp was a trouble maker from Beaver Island and around Christmas he instigated a little rebellion by telling the crew that wages were twice as high elsewhere as they were in our camps. The men fell for his line of talk, quit the camp, and stormed into Metropolitan, where our warehouse and office was located. There they all promptly proceeded to get drunk and threatened to tear down the building and beat up the men in charge. Flannigan and I were in Escanaba and our

man at Metropolitan wired us to come on the first train.
We wired back and told him to hold out, not to pay the men
off and that we'd be there.

We took the first train and arrived about noon the next
day. The men were still drunk and boisterous and we two
took off our coats and went to work. The office and dining
room was upstairs over the warehouse and I met my first
man at the head of the stairs. He was a big, over-grown,
half-breed Indian, drunk and altogether too sassy. I
smashed him under the chin and he went downstairs like a
ton of brick.

"Hey, Flannigan!" I yelled at Larry, who was below.
"Kick hell out of that damned breed!"

"Take care of your own affairs," he shouted back. "I'm
busy."

He was. There were several lumberjacks in front of
him and he was very much occupied in licking the stuffing
out of them all. Some of the men ran a bit too fast for us
but, with those few exceptions, we beat up the whole gang.
They were very quiet when they got sobered up and we took
our time about paying them off. The affair made little difference to us and caused us no inconvenience, as men were
plentiful that season and work was not. We always found
that to be the most effective way of dealing with drunken
lumberjacks. It never pays to let them impose on you. We
stood back of our foremen in everything, kept good order
in the camps, furnished the best of board, and paid the best
of wages. Most men respect and admire a boss who drives
them, as long as he isn't too hard a driver. Some of our men
worked for us continuously for twenty years.

We completed our work for the Eddy-Glynn Lumber Company the third year after starting, 1889-90. We used only one crew on the Whitefish that year and completed the cutting and banking of all the company's timber in that territory. In addition to our work for Eddy-Glynn, we logged eight million feet of timber for the Mann Brothers, of Milwaukee. These men had a woodenware factory at Two Rivers, Wisconsin, known as the Two Rivers Manufacturing Company. Their timber lay along the west branch of the Escanaba River and we drove it down that stream in the spring of 1890. Just about the time we had the logs running well, one of the several dams went out. It had to be rebuilt before the drive could be resumed.

The Two Rivers Manufacturing Company held a charter from the State of Michigan for the improvements on the west branch of the Escanaba and they were supposed to rebuild the dam at their own expense. I wired the company and received a reply telling me to go ahead; they would foot the bill. I filed away the correspondence, certain that there would be trouble before the affair was over, but I didn't have to use it. We drew orders on the Two Rivers Manufacturing Company, found that the banks honored them and promptly flooded the market with them, paying all our bills for reconstruction of the dam and overdrawing our account $2000. When they found out what we were doing, they notified the banks to refuse payment on the orders, but they were too late to save themselves. We had made it necessary for them to look to us for a settlement rather than having us look to them. We settled, but we took our time about it.

When the drive was completed, I went out to collect money from the firms whose logs we had driven. I was in Sturgeon Bay collecting $6000 from the Reindl Brothers Company, when I happened to meet Charles L. Mann, one of the Mann brothers.

"Are Nelligan and Flannigan through drawing orders on the Mann Brothers yet?" he demanded.

I laughed at him.

"I'm all through," I replied, "but the last time I saw Flannigan he was still hard at it."

He was pretty sore and got so loud and boisterous that I finally had to tell him to shut up before I shut him up. Self-preservation is the first law of nature, and we had to follow nature's laws closely in those days. It's always better and safer to trust to your own honesty than to the other fellow's.

In the fall of 1890, Flannigan and I bought ten million feet of standing timber, logged it that winter, drove it the following spring, and sold it to the Menominee Bay Shore Lumber Company of Menominee, Michigan. The same winter we logged four million feet for the Menominee Lumber Company. The next fall we bought twelve million feet of pine along the Sturgeon River for $60,000. We had a fine winter and good water in the spring for driving logs. We sold them delivered in the main Menominee River, for $10.50 per thousand, and cleaned up a fine bunch of money. We also logged four million feet for the Spaulding Lumber Company of Marinette, Wisconsin, the same winter.

Chapter XI   Logging Methods and Camp Life

T HE methods of getting out timber have, of course, been greatly changed and improved since the infant days of the industry. In very early times all the timber was felled by axe men. Two axe men and two sawyers constituted a cutting crew. Then it was seen that sawing down the trees was more efficient. By this practice two men's work was dispensed with. It had taken four good men to fell by axe and saw into logs twenty thousand feet of timber in a day. Two men could equal this daily cut when they used the saw for felling. In some operations today, power drag saws are used instead of hand cross-cut saws.

When I worked in camps in New Brunswick, it was the custom to haul the full length trees to the river bank before they were cut into logs. This was before the days when level, iced logging roads were an essential part of logging operations and it was done because one large trunk could be handled at the expenditure of less labor and time than a number of small logs. When the river was large enough to preclude danger of long logs jamming in the drive, the whole trunk was tumbled in the river without being cut into logs and floated to the booms.

Oxen were used for hauling in the woods almost entirely in the early days, but these slow, steady beasts gave way to horses in the last few decades of the nineteenth century; the horses are now giving way in some places to tractors. Long runner sleds were used with the oxen and, as a general rule, one thousand feet to the load was more than enough.

The oxen had to be shod and this was one of the most difficult chores in the lumber camp, as the ox wouldn't raise its foot like a horse, but had to be lifted from the ground. A heavy frame was built against a tree and the clumsy beast was raised off its feet by wide leather straps passed under its body, fore and aft, and fastened to spools operated by cranks on each side of the frame. In this position, where its feet were accessible to the shoer, it was kept until the job was complete. Due to its cloven hoofs, each ox required eight shoes. The ox drivers were often very brutal and the way in which they handled the stolid, patient beasts was sickening. Goads, sticks about three and one half feet in length with a heavy brad in the end of each, were used to inspire the beasts to greater labor. The drivers would stick these into the brutes until their hides would be full of hard lumps from the harsh treatment.

In writing of oxen, I am reminded of an incident of which I have been told, which occurred back in the sixties, which Delos Washburn, a log jobber from Oconto, was lumbering along the north branch of the Oconto River and, like most loggers of the time, was using oxen in his operations. At the end of each season he would fasten a bell to one of the oxen, so they could be found easily when wanted, and leave them in the woods to take care of themselves for the summer, making a trip to the vicinity about once a month to see that they were all right. One year in the month of June he drove to his camps with a horse team and started rounding up his oxen. He experienced unusual difficulty in locating some of them and, feeling sure that there was no one within miles, he released his pent-up feelings in a lengthy string of choice cuss words. Suddenly a woman's voice rang out from some point across the river, butting in on his profane soliloquy.

"Quit your swearing! Some of the oxen are over here!"

This was deep in the wilderness, about sixty miles from Oconto, where no one, least of all a woman, could be expected to be, and Washburn was a bit in doubt whether the voice came from Heaven or Oshkosh. When he recovered from his surprise and regained his breath, he called back and soon found out that, figuratively speaking, the voice was from Oshkosh. Its possessor was an insane woman. Washburn managed to get her back to Oconto and the papers published articles concerning her discovery. Her relatives came and claimed her. She had been lost about a month before in the neighborhood of Shawano and had been in the wilderness that entire time subsisting on wild berries.

With the improvement of logging roads from mere trails to long, level, icy thoroughfares, oxen gave way to horses weighing about one thousand pounds each. Heavier horses were used and heavier loads hauled as the roads were made better. The largest load of logs ever hauled by one team in our camps—and possibly the largest load of sawlogs ever hauled by a team anywhere in the world—was drawn along an ice road for four miles by a team of horses weighing thirty-two hundred pounds at a camp on the Popple River, tributary of the Menominee, during the winter of 1891. The load scaled 21,603 board feet and was so high that the driver had to stand on the roller of the sled, instead of on top, in order to shun the treetops and keep out of danger if the chains broke under the terrific pressure and allowed the logs to fall off. In order to avoid the possibility of a breakdown on a Monday morning, the load was hauled on Sunday.

The road was perfect. It was along a creek bottom with a slight down-grade and there was a foot of ice on the surface of it. H. W. Cummings of Oconto, who was later killed in a logging camp on the Oconto River, was foreman of the camp. John Barr of Milwaukee, loaded the sled and Dean Ingram of Oconto, an eighteen year old youth spending his first winter in the woods, was the driver.

The intelligence of well trained woods horses is almost uncanny. They can do anything which their bulk and build permit and avoid obstacles and guide the sleigh or travoy (travois) sled with marvelous skill. We used to get most of our draft horses from Iowa and within a month after we put them to work they would be perfectly trained. The work of a skilled woods team with a driver who in the modern parlance "knew his stuff" was a beautiful thing to see.

Although they are exceptional woodsmen in other lines, I always found it wise to keep Frenchmen away from horses. They wear out too many whips. A good teamster must be a man of infinite patience and a Frenchman is rarely so endowed. Some Irishmen make good teamsters, but there is occasionally a bad one like Jim O'Byrne. Jim was driving a thirty-two hundred pound dapple gray team one season in one of our camps on Pine River. Bark had accumulated at the skidways where the teams were loaded, and hauling a loaded sled over the bark was about as hard as hauling it over sand. Street brooms were kept at the skidways for the purpose of sweeping off the bark, but this had been neglected and, when Jim endeavored to get his team under way after the sled had been loaded, he was asking the impossible. The team tried to start the load several times but was unsuccessful. With the sudden, flaring, uncontrollable anger of an Irishman, Jim seized a steel lever which happened to be at hand and struck the left hand horse a terrific blow behind the ear. It fell dead instantly. O'Byrne asked the loader, who had witnessed the affair, to tell the foreman that the horse had fallen dead and they agreed to do so. Shortly after Jim left the camp, the loaders told the truth about the matter. The foreman fired them for not being honest in the first place. O'Byrne could have been and should have been prosecuted.

I always liked horses and, as is usually the case, this liking was warmly returned. When I would enter our pasture, every horse there would race to meet me. I prided myself on being a fairly keen judge of horse flesh, but I was badly fooled once by Sam Newman, a horse dealer of Menominee. A "sweenied" horse is one which has worn out its shoulder muscles from constant contact with the collar and has empty sagging skin where the muscles were. It is useless for work, because it cannot put its weight into the pull without hurting itself. Newman got a sweenied horse once and somehow inflated the empty skin and sealed it up. I looked it over and it seemed in good condition, so I bought it along with some other horses. When it was put in harness and threw its weight against the collar, there were two little blow-outs; I found out I had been stung. It always greatly amused Sam.

The daily routine of life in a lumber camp began long before the break of day. At about four o'clock in the morning the chore boy, awakened by an alarm clock or, more often, by that sixth sense which warns a man that the designated hour of awakening is at hand, would crawl from his cozy nest of warm blankets into the chill early morning atmosphere and start the fires—one in the cook's camp, one in the men's camp, and a third in the camp office, where the foreman and the scaler and perhaps one or two others slept. When a good healthy blaze was roaring in each of the three stoves and waves of warmth were attacking the blanket of cold which lay over the camp like a pall, the chore boy would go into the men's camp and shake the teamsters into wakefulness, being careful not to disturb the sleep of the other men. The chore boy's popularity among the jacks depended largely upon his discretion in this matter. The teamsters would sleepily and noiselessly arise, pull on their outer garments, and depart for the barns, where they fed, cleaned, and harnessed their horses in preparation for the day's work. This done, they returned to the camp, dressed their feet fully, washed for breakfast and, perhaps, took a chew of plug tobacco as an appetizer.

Chewing tobacco reminds me of Ed Erickson. Ed was one of the best woods and river foremen we ever had and he was a gentleman to boot. He started his career as a teamster and he was as good a man at handling horses as he later became at handling men. Like most Scandinavian woodsmen, especially teamsters, Ed loved his chewing tobacco. Whenever he pulled his plug of tobacco out of his pocket, the horses would turn their heads expectantly towards him and he always had to give each of them a chew before putting the plug back. They loved the stuff and Ed, being a gentleman, always treated them, but it ran his tobacco bill pretty high.

By the time the teamsters were ready for breakfast, the camp reveille, blown on a big tin horn, had roused the rest of the camp at about 4:35 A. M., and the jacks had rolled out of their blankets, pulled on the clothes they had taken off the night before—few enough, in truth—taken their heavy socks from the drying racks, donned them and were washing for breakfast. At 4:50 or 5:00 A. M. the "gaberal" would blow the breakfast horn as a signal to the jacks to "come and get it." There would be a rush for the long tables in the cook shanty and a pitched battle would ensue between the lumberjacks and the marvelous products of the cook's culinary efforts, with the jacks invariably the victors. Breakfast in a lumber camp was no such light meal as the morning fruit, cereal, and coffee titbits eaten by modern business men. It was as large and important a meal as any other and the bill of fare would read more like a dinner than a breakfast to the average person of today. Flapjacks or pancakes, sometimes of buckwheat, fried as only a lumber camp cook can fry them, stacked in great piles along the oil cloth covered tables, were favorite items of fare among the jacks. But there might be baked beans, or fried meat and potatoes, or hash, or any other dish which could be prepared from the extensive larder. All this washed down with great draughts of coffee, coffee with such fragrance that one's nose crinkles with remembrance at the thought of it. And there were tasty cookies and cakes. The men were never given an opportunity to complain about the bill of fare in our camps, nor in any other camps for that matter. Lumberjacks were always fed well. They demanded it and it paid the camp operators to feed them well. The better they were fed, the better work they did.

Breakfast over, the men pulled on their outer working clothes and departed for their various posts. Most of them wore wool caps, heavy flannel shirts, mackinaw cloth jackets and pants, heavy German socks and low rubbers. This was the warmest, most comfortable, and most efficient costume for woods work. When the scene of the cutting wasn't too far from the camp, the men returned to the cook shanty for their midday meal, but when it was some distance away, the "flaggin's" were carried to them on a large sled by the chore boys. Great, thick sandwiches, large cans full of hot food from which the jacks filled their tin plates, and great, steaming cans of hot tea satisfied the midday hunger. Back to work they went and labored until after dark. Conditions are somewhat changed now, but in those days there was no eight hour day and while there was day-light the work went on. Then they would straggle into camp and eat their evening meal with appetites such as only tired and hungry men can develop. The teamsters put away and cared for their horses before eating. After supper the jacks would gather around the great red-hot stove in the bunkhouse, pull off their wet, stinking socks and hang them on the drying racks around and above the stove, where they steamed away and emitted an indescribably atrocious odor which permeated the bunkhouse atmosphere.

For several hours the jacks enjoyed themselves to the best of their various abilities. A few, perhaps, read, but there was little to read aside from a few old newspapers and the *Police Gazette,* which was always very popular. In all my experience in logging camps, I remember only one man who ever had a Bible. He was a young fellow spending his first winter in the woods who came of pious parents. They

The lumberjacks are using peavy and cant hooks to move logs to the "A" frame for stacking alongside the loading area on the sled road.

A teamster and skidder stand beside the log pile with their team after finishing the stacking of the huge logs. The logs are now ready for the sled team to come for loading.

A four-horse team sled being loaded along side the skid stack. Notice that the runners of the sled are deep in the runner track. This runner track was kept in constant repair and well-iced so that the sled would ride easily and smoothly through the woods.

whole crew of swampers and skidders working during a light snow fall on the riverside
pile. The piles of logs must be made fairly straight and parallel to the river or the whole
[pil]e will jam when the river ice breaks in the spring floods.

[Th]is picture shows the enormous pile of logs bank and river stacked ready to roll loose at the
[sta]rt of the spring drive. This picture shows hundreds of thousands of feet of lumber which
[re]presents the cutting work of one lumber camp.

The wanigan, or cook shack boat, rests on the bank opposite the site where the lumberme
are cleaning up the logs from the side of the river. There were always some logs strande
during the drive, and river men's job was to rescue the logs with sheer brute force and peav
hooks. Note the river drive pike poles stored on the roof of the wanigan in the foreground.

After the log jam is opened and the main body of logs has passed, the river crew must clea
the banks of the stranded logs which floated to high ground in the high water and were left i
the rush of water when the jam broke.

had given him the Bible when he left home and told him to read it faithfully every Sunday. His intentions were good and he tried it the first Sunday he was in camp, but after watching the lumberjacks enjoy themselves doing the stag dance, the jig dance, and playing games, he put the Bible aside and said, "I'll read it in the spring."

Wherever and whenever men's work is strenuous, their recreation is the same. Reading the Bible wasn't generally considered the sort of thing with which to prepare one's self for another week of hard labor.

"Shuffle the Brogue" was a typical lumberjack game and was often played in the evenings and on Sundays. It was plain horseplay, but it appealed to the lumberjacks and was always productive of a great deal of merriment. A bunch of the jacks would sit on the floor in a ring. In the center of the ring was another jack who was "It." The men in the ring sat close together and passed a rubber around behind their backs, at the same time yelling "Shove! Shove! Shove!" When it was convenient, one of them would hit the man who was "It" in the back with the rubber and· then quickly pass it behind him again and shove it to his neighbor. When "It" caught one of the jacks with the rubber, the caught one had to trade places with "It" and suffer the punishment dealt out by the ring until he caught a man.

Greenhorns in the camp always had to be initiated and this was done in many ways and provided much amusement. One favorite method was the "sheep game." One of the jacks played the part of a farmer who owned a sheep, another posed as a sheep buyer and the greenhorn was rolled up tightly in a heavy blanket and became the "sheep." He was carried by two other jacks. The farmer and the sheep buyer would stage an argument over the weight of the sheep. To determine its real weight they would let it down repeatedly on the "scales." The "scales" was a sharply pointed stick and the "sheep" was always thrown onto the "scales" in such a way that the point of the stick came into violent contact with the tender, rear central portion of his anatomy. This was very uncomfortable for the greenhorn and very laughable for the rest of the crew. Another favorite stunt was to send a greenhorn to the cook shanty to borrow the "bean hole."

The relation of the feats of Paul Bunyan was always a pleasant way of passing the time. There were usually one or two members of the crew who were familiar with all the tried and true versions of the mighty Paul's feats and who had sufficient imagination to add a few original tales of their own. They would tell of how Paul logged off the Dakotas by hauling a section of land at a time to the landing with the help of Babe, the Big Blue Ox; of Johnny Inkslinger and of how he saved nine barrels of ink one winter by leaving the dots off his "i's" and the crosses off his "t's"; of the Little Chore Boy, who had to turn a grindstone so big that it was payday everytime the handle came around; of the Seven Men and of how, before the grindstone was invented, they sharpened their axes by rolling rocks down hill and racing along side them while they held the axe edge against the rolling stone; of Babe, the Big Blue Ox, which was so powerful it could pull a crooked road straight, but wouldn't work on a dirt road in the spring, so the road had to be whitewashed to fool it; of Big Joe, the cook, who had to keep digging a hole in the deep snow drifts for his flue all winter long and who, when spring came, found he had a hole one-hundred and seventy-eight feet high standing in the air above his chimney. These impossible tales were all told with the utmost gravity, and the greenhorns would sit and drink them in with open mouths until they came to a realization of their absurdity. Then they would find them uproariously laughable, but no old timer ever cracked a smile while telling or listening to one. It was the unwritten law to remain serious.

Such sessions, of course, were only occasional and most of the time the evening was taken up with shop talk, rough banter, stories, feats of physical prowess, and games. After a few hours of this, the smoky oil lamps would be blown out and the tired men would roll into their bunks. Quiet reigned for a time and was then broken by a chorus of hearty, healthy, masculine snores which steadily increased in volume until it rivaled the sough of the night wind in the pines outside.

The bunks were usually double decked and lined the walls of the bunkhouse. Some camps had mattresses while others simply had bunks filled with clean straw. A funny thing happened in one of the Holt Lumber Company's camps one year. A straw from an upper deck bunk fell into the mouth of a lumberjack lying in the bunk below and lodged there. He had a great deal of trouble with it and had to have medical attention to have it removed. He sued the company for damages and, if I remember correctly, he received something .

Camps were none too clean, due to the roving habits of the men. Today it is easier to keep a steady crew of men. Henry Ford, I understand, has greatly improved the conditions in his logging camps and has set a splendid example for others. Clean sheets and pillow slips are distributed every Saturday. That was out of the question in the old days when there were no bathing facilities in the camps and men roving from one camp to another would louse the very horses in the barns. But we always kept clean quarters for ourselves and if a clean stranger came to the camp, we had a clean bed for him without extra charge.

Mr. Ford has also, I understand, provided ample bathing facilities for his men in the camps and made a rule that every jack must bathe at least once a week or leave the camp. In the old days most of the jacks went without bathing all winter long and even when spring came some of them were loath to indulge. Flannigan once got a bunch of them drunk after the camps closed in the spring and took them into a bath-house where he ripped off all their clothes and made them bathe. It was about the only way such a thing could have been accomplished. The lumberjacks washed their faces and hands morning and evening during the winter and bathed their feet at intervals to keep them in good condition. Some of them shaved, but most of them went with long beards during the winter months. They cut each other's hair when it became inconveniently long.

On Thanksgiving Day once, in one of our camps, somebody who was not very well informed on the matter put the question:

"What in hell is Thanksgiving Day, anyway?"

Tom Conlin, an Irishman from Massachusetts, who was a staunch Catholic and couldn't quite see the reason for a holiday which was not a Holy Day, replied:

"I'm damned if I know. It's some kind of a Protestant day, I guess."

Whatever kind of a day it was considered, we always made it a point to celebrate it in our camps with a big turkey dinner, although it was not customary to lay off work. Turkeys were always sent to the homes of married men employed in the camps. This was also done on Christmas Day. There was no work on Christmas and sometimes some of the men left the camp and went home for a day or so if their homes were not too far distant. Richard Doyle, "The Scout," our noted foreman, tried to make the men in his camp work one Christmas Day, but he met with mutiny. The men refused and went to Iron Mountain where they hired a lawyer to collect their wages for them. The lawyer somehow got service on Flannigan and we paid their wages without making any trouble. They should have come to us first and they would not have had to pay attorney's fees. The work was never interrupted on New Year's Day, although there was sometimes a slight celebration in the form of an especially large dinner.

Catholic Sisters often came to our camps collecting money for orphanages and one could not help but admire their courage in going alone into the wilderness among the roughest of men without the slightest kind of protection. It is to the credit of those rough men that the sisters were almost invariably treated with courtesy and accorded the respect due them. Some there were, of course, who were always asking: "What business have they in the woods?" But most of the lumberjacks respected them and their mission and gave freely and gladly to help them. It was through the humanitarian efforts of these women, they were well aware, that their illegitimate offspring, the unintended results of their wild revels, were reared in decency and given a chance in the world.

When they came to our camps, we always used them kindly and did our best to make them comfortable. We extended the same courtesy to the women from the Good Will Farm at Houghton, Michigan. A couple of them came to one of our camps once and the foreman, in accordance with our usual custom, turned the camp office over to them, supplied them with towels, gave them kindling wood for starting their fire in the morning, and furnished what other meager comforts he could. They spent the night there. The following morning, when they did not get up at the usual early rising hour of the lumberjacks, the miserable cur who was acting as cook passed some insinuating remarks about them. The crew paid no attention at the time, for they were eating, and eating is a thing of paramount importance in a lumber camp. But when breakfast was over, they rose in a body and went after the cook. They wanted to lynch him, but he was a fast runner and he got away. For all their rough edges, the lumberjacks were a chivalrous breed and the man who dared to be careless in his comments on respectable women was taking chances.

The sisters took terrific chances in their wanderings. Fred M. Stephenson, driving into one of his camps in the wilderness along the Fence River once, found two of them huddled over a tiny, inadequate fire of twigs which they had broken off the trees. They were lost and the thermometer was then at ten degrees below zero and dropping lower. They were almost frozen and would probably have perished in a short time had not Mr. Stephenson happened upon them. He took them into one of his camps and provided for their comfort.

During the nineties we had a logging camp at Carney, about eight miles north of Pembine Junction of the Chicago, Milwaukee and St. Paul Railway. Two sisters had been at the camp collecting money for an orphanage in Chicago and were waiting to catch the train into Pembine Junction. I had also been at the camp and wanted to catch the train to Pembine, so I could make connections with the Soo Line train going west to Crandon. The northbound trains usually stopped at the station to let off men bound for the camp, and the night train going south would stop if they were not behind schedule and if they were flagged. I had the camp chore boy come down to the track and flag the train with a lantern. It was evidently behind schedule for the engineer paid no attention to the signal and the train roared through like a bat out of hell. I was pretty sore and, as I learned later, gave vent to my anger in a few well-chosen words of invective.

The camp office was only about twenty rods from the track, so I told the sisters to go and occupy my bed and I walked the eight miles into Pembine in two hours, carrying a lantern. When I arrived there, I was still seeing red, so I wrote a message to Mr. Menteran, superintendent of the Green Bay division of the road, and in it told him the story with excusable and colorful exaggerations, making it sound pretty bad. It was effective, for after that all trains got orders to stop when flagged and they obeyed them, and we had no more trouble. The chore boy told me I swore terribly when the train went through without stopping and perhaps I did, I don't just remember now. But the circumstances justified it even if I did offend the sisters' ears a bit. I gave them my bed and did the best I could for them, so I guess it was all right.

We used to encounter some strange characters in the woods, men far different from the average lumberjack. One such was a young Englishman who came to one of our camps sometime in the nineties along with a few jacks from Milwaukee. He didn't know how to handle an axe so "The Scout," Richard Doyle, who happened to be foreman of that particular camp, put him to work with the road crew. His hands were very soft and unused to such labor and in a few days they swelled up and blistered, but he gamely stuck to the job. I had noticed his hands and the next time I happened to be in the camp, a few days later, I saw that they had not improved but were getting worse. I had "The Scout" send him to me at the camp office. He told me how he happened to be there and a bit of his personal history. He was a young Englishman of fairly good family who had had some trouble with his folks and cleared out. He drifted to this country and finally landed in Milwaukee. In search of work, he fell in the clutches of an employment agent and was shipped off to the woods with a bunch of jacks. He seemed to be a very nice, young fellow and fairly well educated, so I gave him a light job at keeping the men's time and the camp accounts so that his hands would have an opportunity to heal. They got better and, as soon as he became well-acquainted with the men, he got along fine and was well-liked. He was a good clean liver, and never used tobacco or touched liquor. He stayed with us until the camp broke up in the spring. Occasionally he wrote to some of his people and the answers always came in our care, of course. The next summer his folks wrote us inquiring as to his whereabouts. I lost track of him and never heard of him afterwards.

PART III

WHAT THE WOODS SCHOOL TEACHES

Chapter XII   Dramas of the Drives

In the old logging days of Wisconsin and Michigan, every spring saw the curtain roll up on a tremendous drama along the rivers of the timber country; a drama greater even than that in which the giant pines were felled and dissected; the epic drama of the drive. All winter long preparations went forward for this brief period of intense activity and struggle. Thousands and thousands of logs were banked along the riversides, or on the ice of the streams. There they lay, in serried ranks, awaiting the day when they would be tumbled into the streams and rivers, to become the multitudinous parts of a mighty, surging monster, the drive. The sun, in its daily journey across the heavens, worked ever northward, each day adding a few moments to the time taken by that journey, each day increasing the heat by which ice and snow were changed to water. Hoary old King Winter died and a new monarch, Princess Spring, mounted the throne of the earth. In reservoir dams at the headwaters of the rivers the waters of the spring thaw, results of the sun's sovereign alchemy, were stored up to carry the logs along the first lap of their journey to the mills.

In the camps cutting operations came to an end. Some of the men left for an early spring spree. Others stayed on, changing their rubbers for calked boots, the many spikes of which, on sole and heel, were filed to sharp points that would bite into the pine logs. The heavy flannels of winter gave way to overalls which would be lighter when wet and would dry out faster. These rivermen were the pick of the camps, lumberjacks of unusual strength, agility, daring, and hardihood. They had to be. For days they had to go with but little sleep and with snacks of food snatched whenever and wherever possible. They had to suffer frequent duckings and were almost continually soaked to their skins at a time of year when the weather was still far from clement. The price of their safety was constant and unfaltering vigilance. They worked in a treacherous element, and the slightest misstep or miscalculation might send them relentlessly to their deaths.

These men were of all nationalities and creeds. Frenchmen and Irishmen, English and Scotchmen, Indians and Scandinavians. The latter were always good rivermen, but for some unknown reason there were relatively few of them on the rivers, although they were numerous in the camps. The rivermen wore their overalls stagged, that is, they were cut off at a point just above the boot tops. This cutting was usually done with a knife or ax and was, as a result very uneven. The rivermen took pride in the irregularity of the stagging and they usually cut one leg shorter than the other. Flannel shirts adorned the upper parts of their bodies as a rule, and the most nondescript hats imaginable rested on their heads. They carried peavies or pike poles, with which they wrestled with and directed the course of the logs along the wandering rivers. Such were the American rivermen, picturesque and heroic figures, in truth. Hard-living, hard-drinking, hard-fighting, blasphemous pioneers who have gone the way of our other typically American pioneers, the frontiersmen and the cowboys, and are now nothing more than a tradition.

There would come a day in spring when the gates of the reservoir dam on each stream were lifted and its stored up waters turned loose. Down the river bed they rushed, eager and irresistible, tearing up and carrying along the rotten ice which covered the stream. The rollways were broken out and, in a smother of spray and a tumult of noise and confusion, the logs were tumbled into the rushing waters. Out upon their heaving, surging backs scampered the rivermen, pushing, pulling, and prying with their peavies and pike poles, doing their best to keep the logs always on the move. The drive was on!

Down the river it wended its way, around bends, over falls, through rapids. Upon the constantly shifting carpet of logs the agile rivermen labored in a Herculean manner. Whenever there became discernible the slightest cessation in the steady, downstream movement of the logs, a jack was on the job to remedy the matter. Every effort was made to avoid jams, which were apt to exact a heavy toll in both men and money. The first lap of the journey came to an end at the dam next below the reservoir dam. In the great pond back of this dam the logs came to a temporary rest, while a sufficient head of water was raised to carry them along the next lap.

During the day there was usually an upstream wind and this drove the logs to the back end of the dam pond. This made it necessary to work them down at night, when there was little or no wind, and to throw a boom across the pond to hold them close, to the sluiceway. When a good head of water had been raised, the gates of the dam were pulled up and the water allowed to run through for some time before the logs were sluiced. This is due to the fact that the logs will outrun the water and strand themselves in a jam if they are released soon after the gates are opened. The length of the interval between the opening of the gates and the sluicing of the logs was usually decided by a calculation of the speed of the current, determined by throwing a board in the water and timing it for a mile, and the distance between the dam and the next one below.

Sluicing the logs, that is, directing them through the gates of the dam and down the sluiceways, is rather dangerous work. If a man happens to be washed into the sluice, there isn't much hope for him, as there are too many swiftly moving logs around to knock him senseless. Once when I was working with "The Scout" and some other men at the head of a sluiceway, a long Frenchman who must have been about a foot taller than I was, made a misstep and fell into the water. The current was rapid and strong and the Frenchman was quickly swept toward the head of the sluice. Doyle was closer to him than I was, so I raised my voice above the roar of the water and yelled: "Hey, Doyle, save that man!"

"To hell with him," replied Doyle, with characteristic abandon. "He's no good anyway."

So I had to go after the Frenchman myself. I managed to pull him to safety, but almost got sluiced myself in the bargain.

In 1880 or 1881, when we were lumbering on the Peshtigo River and driving the logs out of the middle inlet of Lake Nocquebay, an Irishman from Canada named John

Powers was working below the gates of the dam. He was smoking a short, typically Irish, clay pipe and his work was to keep the logs from jamming as they came out of the sluice-way. Somehow or other he made a miss-move, fell into the water, which was rather deep at that point, and disappeared beneath the surface. A man was standing above on top of the dam and saw the accident. Immediately, with the quick wit and action which are essential parts of the riverman's equipment, he dropped the gates of the dam. The water fell instantly and revealed Powers calmly standing on the river-bed, his short pipe still clamped firmly between his jaws. Neither he nor the pipe were any worse for the wetting.

When, in spite of the unfaltering vigilance and quick action of the rivermen, the logs piled up into jams, there were enacted the most interesting scenes of the drive. Thousands of logs would sometimes work themselves into an intricate mass, as hard to take apart as a Chinese puzzle, and hold in check tons of angry, pent-up water. There were many ways of breaking jams and the method used usually depended on the size of the jam and its general condition. A small jam could often be broken up by the rivermen with their peavies. They would attack the logs on the side of the stream where the current was swiftest and push and pull the logs into the current one by one until the whole jam was broken loose. As a general rule there were one or two key logs which held the jam and when these were pulled out, the jam would break up or move on to another position, in which case the removal of another key log or so was necessary. The riverman must be constantly alert, for when the jam breaks, it usually goes out in a hurry and he must run for his life over the surging logs. Sometimes a key log was removed with a swamp hook. This was in cases where the key log was wedged very tight, or where it was unusually dangerous for the men to work. The swamp hook was placed on the key log and its teeth made to bite deep by the use of a trip line from one side of the stream while the gang of men pulled a rope from the other side and endeavored to extricate the log from its position. In Canada I've seen a line stretched taut over a jam and a man pulled out on this line and lowered in a harness to locate and dislodge the key log. Dynamite is by far the safest and most effective method for breaking up large jams.

There were always certain bad spots along a river at which the logs were likely to get tangled up. Sturgeon Falls on the main Menominee River was one of these. Early one spring when we were driving the Menominee, we had a jam of ice and logs just above the falls. Just as the jam broke, a drunken Indian ran out on the ice and logs, for no reason at all, and was swept over the falls in a grinding, crashing mass of ice cakes and great sticks of timber. There was, of course, no doubt in our minds about the Indian's fate, for we were sure that no human being could live in that inferno of ice and wood and water. We didn't pay much attention to the incident, as a drunken Indian more or less really didn't matter much. Once again, as in the "Paddy" incident at Oconto Falls, which I have related in a previous chapter, we were given a lesson in the resiliency and hardihood of the human body. Some time after the Indian had taken his cold plunge, he reappeared on the scene, much as "Paddy" had. He was well sobered up, but his clothes were torn to shreds and his cap, like "Paddy's" hat, had disappeared.

There were sometimes small operators along the streams who did everything in their power to get their logs down to their mills at the minimum of cost, even going so far as to throw in their logs with another drive. When Flannigan and I were driving twelve million feet of timber out of the east branch of the Sturgeon, we had an experience with such a fellow. He had built a small sawmill and dam at Loretta on the main Sturgeon River where the Chicago and Northwestern Railroad bridge crossed the stream. His logs, a rather small cut, were banked some distance upstream and when our drive went by, he had his men break out his rollways and put his logs in with ours so that they would get downstream without any cost to him. Although unfair, this was perfectly legal and we couldn't stop him.

But when we got down to his dam, he tried to hold up our drive while he sorted out his logs and divided them from ours. That was asking a little too much and we paid no attention to him. We put our men at the sluiceway and ran through all the logs, both his and ours, as fast as possible. Of course, he lost all his logs; I don't know whether he ever got any credit from the boom company for them or not. If he'd have come to us in the first place and asked for help, we'd have been glad to assist him and would have done it free of charge, but we didn't like such high handed tactics. He said he would sooner lose the logs than have my blasphemous crew around sorting them out. I wasn't there at the time, but I suppose those who did bother him were some of my students.

Grilling work coupled with such hardship and exposure as the rivermen were forced to stand invited the use of liquor and whenever possible the men obtained it. When used in

moderate quantities, whiskey is a helpful stimulant for men
engaged in such work, but excessive indulgence is dangerous.
Men under the undue influence of liquor cannot drive logs
without great peril to life and limb. It befuddles a man's
mind, clouds his eyes and makes unsteady his feet, and safety
and continued existence upon the logs depend upon the sure
functioning of mind, eyes, and feet. But men are prone to
forget those things and, when the opportunity offered, the
rivermen were never loath to get themselves soused.

We were driving the Sturgeon River one season and
camped at a dam about two and one half miles from Metropolitan, an almost abandoned mining village. As usual, we
had to wait until sufficient water was stored up to carry the
logs down the next lap of the river. The men wandered off
to the village and proceeded to get gloriously and uproariously drunk. They straggled back to camp and were just
about sobered up and ready to resume the drive when Swan
Anderson, a Swede saloonkeeper, appeared on the scene
with a jug of whiskey. The men got drunk all over again
and we had to hold up the drive for another day. I was
pretty sore and I dug up a copy of the *Life of the James
and Younger Boys* somewhere and found Quantrell's oath
in it. This oath was a bad one and I made the jacks repeat
it after me word for word, in all picturesque profanity, their
hats off and their right hands up, and finally solemnly swear
that they would abstain from Swan Anderson's whiskey
and beer until the drive was completed. . They kept the oath,
but only because it was impossible to break it, as there was
no more of Swan Anderson's liquor readily available until
the drive was over.

A riverman's dream came true one spring when we were driving on the Nett River. A freight and passenger train on the Chicago, Milwaukee and St. Paul running from Channing, Michigan, to Ontonagon was late and the engineer, speeding to make up lost time, wrecked the train on the Nett River bridge. A box car full of cased whiskey was broken open and, when our men found it, the liquor magically disappeared. They hid it all through the woods and they themselves were scattered all around, dead drunk. I wired the division superintendent at Green Bay about it and he sent a bunch of section foremen to find the whiskey. But most of them were worse than our lumberjacks. Our drive was held up for a week while the bridge was repaired and by that time almost everybody was sober again.

A driving crew was divided into two sections, one of which handled the fore part of the drive, breaking up jams, sluicing, and so forth, and the other the rear end, picking up logs which were stranded when the high water receded. Although the jam crew had a bit more dangerous work, the members of the rear crew had to labor much more strenuously and they were paid accordingly. On long drives the "wannigan," a raft on which the cook carried supplies and utensils, came down with the "rear" and whenever possible the men ate regular meals. When this was not possible the men carried what were called "nosebags," haversacks filled with lunches. Canned tomatoes, eaten cold out of the tin, were a favorite dish with the rivermen, as they were very refreshing on warm spring days. Tea was usually used instead of coffee, because it was a better stimulant and more refreshing.

Only pine logs could be driven with real success on the rivers. A large proportion of hemlock, basswood, and hardwood logs were apt to sink and so they were usually railed to the mills. There must be over one hundred million feet of sunken timber in the main Menominee River and its tributaries which have been there for the last twenty-five years. When the logs of a number of different companies were all takn down in one drive, as was often the case, the drive came to a stop in the sorting booms and there the logs were divided according to the marks stamped on their ends and taken separately to their respective mills.

Stray logs, that is logs which were not end marked or bark marked, were taken care of by a committee appointed by the boom corporation. They were sold and the boom company, which was often jointly owned by the millowners and loggers, received the money.

The mills which received the cuts of independent loggers always scaled the logs in the woods and almost invariably they had their scalers instructed to give low scale and thus cheat the loggers. Of course, there were some honest mill owners who did not stoop to such practices.

Occasionally, at camps along the river or at the mill towns where the drive came to an end, the rivermen would indulge in a bit of sport on the logs. Two or sometimes more would jump on a log together and roll and snub it furiously to see which could stay on longest. Sometimes they all took a ducking, sometimes one of them proved himself better than his comrades by staying atop the spinning timber and forcing it into subjection. Out of this grew the modern sport of birling or log rolling, which is rapidly attaining national popularity. It is a fast, verile sport which demands of those who take part in it speed and endurance to a superlative degree.

Death constantly walked by the side of men on the river and made its frequent appearances in the most casual and unexpected ways. It was so casual that it was treated almost callously. A drowning did not even slow up the work. The drive had to go on. No mere human misfortune or agency could stop it. Not even death could stop it. A slight miscalculation, a slight misstep and a man might disappear beneath the logs on which he had worked. The body was recovered at once if possible. The work went on. Jaws set a little harder, eyes blinked away moisture which might distort vision and prove fatal, feet stepped a little more carefully, and that was all.

An incident occurred on the west branch of the Fence River once which was typical of the casualness with which death struck. There was a jam and we were eating lunch about a mile upstream from it. Lunch over, we were to go downstream and work at breaking up the jam. The water was pretty rapid and the men were talking about riding logs in it. Anything which savored of unusual danger was always tempting to them.

"I guess I'll ride a log down to the tail of the jam rather than walk," announced one of the crew, a man named Bell.

A riverman on the drive hates to walk as much as a cowboy is reputed to. Bell jumped on the log and started on his journey downstream, his peavy in his hands. As he approached the tail of the jam, his log struck a rock and he was thrown into the irresistible current and swept like a stick of driftwood to a death beneath the jam. After the jam was broken, we recovered his body and sent it out to Crystal Falls.

But even grim tragedy had its laughable side. We were driving the Fence River one spring and were almost out of the main Fence when a young fellow from Iron Mountain ran out on the logs in the middle of the stream, where there was no work to be done and he had no right to be. He became frightened, lost his footing, fell in, and was drowned. The flood of water from the dam above was almost spent and about an hour after he went down, we found his body in the Golden Rapids. We wanted to get the corpse out to the station at Witbeck that night. Our four mule supply team was at Camp One so we put the corpse in the wagon. There were six men in the camp who were cripples. Some of them had sprained ankles and others had run picks through their feet. They also had to be taken to the station, so they piled into the back of the wagon with the corpse. In addition to the corpse and the cripples we loaded the wagon with six boxes of dynamite which had to be taken to the railroad, fifty pounds to the box, three hundred pounds in all. This completed the load and the team started out for Witbeck, fifteen miles distant.

The night was dark as pitch, the menacing growl of thunder filled the air, there were frequent flashes of lightening and the rain descended literally in sheets. We had covered the corpse with a couple of blankets, but the crippled lumberjacks logically decided that it was more important that live men be comfortable than that a dead man be kept dry, so they pulled the blankets off the body and covered themselves. When the team finally arrived at the station, the supply teamster heaved a great sigh of relief.

"The worst blankety-blank load I ever hauled," he exclaimed, with expressive profanity, "a dead man, six cripples, and six boxes of dynamite. And I could have lighted my pipe with the lightnin' anytime!"

In the summer of 1892, Lawrence Flannigan, Fred Carney, Jr., and I entered into a triple partnership and organized the Fence River Logging Company. Mr. Carney acted as secretary and treasurer. Mr. Flannigan as vice-president and myself as president. For our first job we signed a contract to log and drive one hundred forty million feet of timber along the Fence River in the upper peninsula of Michigan for the H. Witbeck Company. The logs were to be delivered in the main Menominee River and we were, of course, allowed several years for the completion of the contract.

The Fence River is a tributary of the Michigamme River, which in turn flows into the Menominee. It derived its name from the fact that the Indians used to build brush fences along it with which to trap deer in the early days. They would cut down small trees and brush and make fences ten or twelve miles in length at right angles to the river on both sides. About every eighty rods a gap was left in the fence and there a scaffold or roost was constructed. At these points of vantage the Indians waited for their prey. When the deer, in their movements up and down the river, came to the fence they would follow it to the closest gap and the Indians would shoot them as they passed through. This hunting was done mostly in the fall and spring, during the semi-annual migration of the deer. Each fall they journeyed south to get away from the deep snow and wolves, and each spring they went north to spend the summer.

On the first of September we started our work along the Fence, cutting supply roads and building camps and barns in preparation for the cutting of twenty million feet of timber, the amount we intended to manufacture the first season. We had four camps that winter at various points along the Fence and our office and supply house was located at Witbeck, Michigan, which was named for the Witbeck Lumber Company of Marinette, Wisconsin, of which Daniel Wells, Jr., of Milwaukee, J. H. Witbeck of Chicago, and Frederick Carney, Sr., of Marinette, were the owners and managers. Our foremen in the camps that winter were George Moore, popularly known as "The Bear," Richard Doyle, known as "The Scout," M. J. or "Mike" Flannigan, Larry's brother, and Richard Hayes. They were all good foremen and they achieved results. By the end of March, 1893, our twenty million feet were in rollways at the banking ground all ready for the opening of the spring drive. Operations in the camps were suspended and the men came out to Witbeck to get their checks and a five dollar bill apiece to take them into Iron Mountain or Marinette, where they could get their checks cashed and embark with unrestrained enthusiasm on their annual spree. None of them had had haircuts all winter long and they looked like members of the House of David. On arriving in Marinette or Iron Mountain they usually bought new clothes, had their hair cut and their faces shaved, but few of them ever took a bath.

During the month between the suspension of cutting and the beginning of the drive, which opened the first of May, we kept a small crew of men at work along the river fixing the driving trails, blasting rocks out of the channel and in other ways preparing the stream for the advent of the drive. The snow had been pretty deep that winter so there was plenty of water for the drive in the spring. It held up well all the time we were running the logs down the Fence and the Michigamme into the Menominee.

There was one man drowned on the Fence that first spring, a young fellow named John Heger. We didn't find his body until June and when we did the remains were in such shape that we couldn't give him a decent burial, but had to inter him in a knoll on the bank of the river. We notified his father at Cheboygan, Michigan, who didn't seem greatly concerned over the matter. We sent him a check covering the time the young man had worked for us.

The drive was brought to an end around the middle of June about four miles from Florence, Wisconsin. That done we shipped our bedding into Witbeck to be cleaned for the following season and then did little more than sit around and look after the horses until September, when we started on our second season's work.

Deaths in the camps were not so common as on the drive, but they did occur occasionally. Once a man was killed in one of our camps and I took the remains out to Republic, Michigan, where they were prepared for burial before being sent to his people. The undertaker in charge was a Swede named Carl Peterson. He took me into the room where he was working on the body and then walked out and locked the door behind him. The odor was not exactly pleasant and I had no inclination to stay in that room.

"Lemme out of here," I yelled at Peterson, "or I'll kick the devil out of everything in the room!"

The fool Swede only laughed.

"The first time in my life I ever had an Irishman where I wanted him," he shouted back through the door.

He finally let me out and we both laughed over it. Peterson was a good fellow and we were the best of friends.

We started operations on the first of October, 1893, and opened three more camps, making a total of seven. That second year we cut, banked, and drove about thirty million feet of pine logs for the Witbeck Company and others. We again enlarged the scope of our operations the third winter. We had eight large camps along the Fence, Nett, and Pine rivers, and we worked two hundred of the finest horses money could buy. There had been much extravagance among the camp cooks, so I offered a prize of twenty dollars to that one of the eight who could keep his kitchen the cleanest and board his men the best and cheapest. The scheme worked to perfection. They all buckled in with great enthusiasm and at the end of the season they had all done so well that I had to give each of them a twenty dollar bill. It was money well used. They must have saved us at least a thousand dollars.

We landed five million feet of logs on the ice on Fence Lake on the east branch of the stream and when spring came, we found ourselves in a bad fix. The Fence River Improvement Company had built a dam at the outlet of the lake but there was no inlet, and although the dam was kept closed for six months, the water did not rise a foot. We had signed a contract to deliver the logs in the main Menominee River and the situation was desperate. It called for unusual action so I took my compass and set out for a good look at the surrounding ground. It didn't take long to find a practical solution. The Spruce River, a tributary of the Michigamme, was separated from Fence Lake by a hog's back hill. I had a crew of men throw a rough dam across the Spruce and then cut a canal through the hogback at a point back of the dam. We made Spruce River, or at least a part of it, run into Fence Lake. At the time I cut the canal I didn't even know whose land I was trespassing on, but there was no time to waste in worry about such things. After we got the water running, we found that the land was owned by the Cleveland Cliff Iron and Land Company, which was very reasonable about the matter and let us use the canal at a rent of one hundred dollars per year from that time on.

Richard Doyle, or "The Scout," as he was commonly called by all of us, took charge of Camp One on the Pine River when we started operations there. In his prime he was one of the most capable logging foremen in the country; I have never known a better one. Like myself, he was raised in New Brunswick and came west as a young man. It was not long before he was thoroughly versed in the practical aspects of logging and given an opportunity to prove his unusual capability as a leader of men in the position of foreman. His exceptional command of expressive profanity did much to inspire the respect of the men who worked under him and, when it became necessary, he could command a more healthy respect with his fists. When he first bossed a camp for us, I located him at the corners of the section we were to lumber, gave him a map of the town and range, and showed him with a compass the degrees the lines were run on by the government surveyors. He learned it all very quickly, understood it thoroughly, and was very apt at picking up the other technical knowledge it was necessary for a foreman to have.

Like the lives of so many other wild woodsmen of those hectic years, Doyle's life was blighted by an insatiable thirst for whiskey. If liquor was unobtainable, he would work faithfully and soberly the whole winter through, but when spring came and the camps closed down, he would collect his winter's wages, hot-foot it for the nearest street of saloons, and squander every dollar of his hard earned money in one grand spree. About the time he was fully sober again, the drive would start. He would go on that and then, with another roll burning a hole in his pocket and his vitals burning for the taste of raw whiskey, he would embark on another drunk.

Year after year the performance was repeated. Once, with his consent, I kept three hundred and fifty dollars out of his pay, banked it for him, and showed him the certificate of deposit. He was satisfied for the moment but the damned saloon keepers, who hated to see all that money slip through their fingers, kept telling him he'd lose it and he pestered me for it until, in despair, I gave it back to him. Upon which he promptly went out and blew it all on another spree. He started east once, a good sized stake in his pocket, to visit his mother in lower Canada. But he stopped over too long in Montreal, spent every last cent of his roll, wired us for money, and came home without ever seeing his mother, although he had traveled a thousand miles with that intention, and was within a relatively short distance of her when he had to return west. In his latter years the whiskey got the best of him, as it does of most heavy drinkers. His stomach was burned out to the point where it wouldn't hold any food and his health gave out completely. After working for us twenty-five years, he went west to the state of Washington, and from there to British Columbia, where he died.

Although ordinarily a model of industry and energy, Doyle was hardly to be relied upon when there was any drink of an intoxicating nature within smelling distance. On one of my frequent trips to the Pine River Camps, I stopped at our warehouse and office at North Crandon and in running through some orders from Doyle's camp I discovered a requisition for two gallons of alcohol. I asked the clerk about it and he said that he had been sending it in for some time at intervals of about a week; that the teamsters wanted it for their horses. It sounded like blarney to me and I had a hunch that the teamsters were getting more of the alcohol than the teams, so I told the clerk not to send any more in until he heard from me.

I went to Camp One that day and early the next morning about the time the teamsters should have had their horses fed, cleaned, and harnessed, I walked into the barns and found a convivial little gathering, indeed. The men had a kettle of hot water, some sugar, some tin cups, and some spoons. And with the alcohol which was supposed to be for the horses they were busily engaged in making and consuming hot slings, or toddies. They were all pretty well lit up and Doyle, the devil, was the life of the party. They all wanted me to join them, but I shut down hard on that and made them get to work. I stayed at the camp all that day and there were more sawlogs hauled while I was there than there had been any day for a week. They had been too busy making hot drinks to think of work.

Hard drink interfered with the work all too often. We always bought railroad tickets for the lumberjacks so they could get into the camps in the fall; when we lumbered on the Pine River, they would go from Marinette to Pembine Junction by the Chicago, Milwaukee and St. Paul Railroad and from Pembine Junction to North Crandon on the Soo Line. At Pembine Junction they would have the ends of their tickets left for the trip to North Crandon. These ticket ends were worth about two dollars and a half apiece and the first thing some of the jacks did when they arrived at Pembine Junction was to go into the first saloon they saw and cash them for a drink or so of whiskey. I redeemed many of them and even had a few jacks put in jail for the offense. But it did no good.

Quite often the love of the lumberjacks for liquor resulted tragically. One winter when we logged along the Nett River, five men left one of our camps on a Saturday evening. They arrived at our warehouse, on the Ontonagon branch of the St. Paul road, too late to catch the Saturday night train. There wasn't another train until Monday evening, so the quintet settled up at the office and, well-stocked with money, set out for Amasa, eighteen miles away. There they got royally drunk and started on for Channing, Michigan. As the effect of the liquor wore away, they got drowsy and all five of them lay down on the railroad tracks and went to sleep. A freight train roared over them and killed every one.

The timber was very plentiful along the Pine River. In at least one section near Camp One, it stood about one million feet to every forty acres, or sixteen million to the square mile. The scene of the cutting was quite a distance from the banking ground or landing and the camp buildings were situated alongside the main road about half way between the two. We had a crew of eighty-five men and twelve teams of horses in Camp One the first year we operated on the Pine River and the total cut was about six million. When the logs were all cut, piled high on the skidways and the ice roads in first class shape, Doyle began to haul his six million. The sprinklers were kept running all night long, every night, while the hauling was in progress and the road was a perfect sheet of ice from woods to landing. All the teams would be in camp every night, half of them bound for the landing with full sleds, the other half bound for the woods with empty sleds.

Camp Two, at Farrell Dam on Pine River, was bossed by James Holmes of Menominee, Michigan. There were fifty men and eight fine teams of horses employed at that camp. That is, the horses were there until they were cremated one night in a barn fire started by some rotten incendiary. In those days there were always a few hangers-on around every lumber camp, men who lived in the woods in rough shacks and eked out an existence by trapping, pot-hunting, and stealing provisions from the camps. Several of these were living near Camp Two, and Holmes had had trouble with one of them, known as "Bay Shore Mike." There was a watchman who was supposed to keep a strict guard over the camp buildings during the night, but he went into the cook's camp and went to sleep one evening when he should have been on the job. Out of spite for Holmes, one of the loafers, probably "Bay Shore Mike," sneaked into the camp and set the barn afire. The blaze was well started before it was discovered and all of the sixteen horses were lost. Those which weren't killed by the flame, we had to shoot to put them out of their agony. The man who was supposed to be watching over the camp and was collecting wages for the performance of that duty, cleared out, and no one, so far as I know, saw him from that day forth. He was probably a friend of those confounded loafers. Devils who would stoop so low as to set fire to a barn full of horses would not hesitate to burn up a men's camp either. The worst of it was we always tried to treat them well. If a tramp stopped at one of our camps over night, he was always sure of a meal and a bed.

Once, while he was foreman of Camp One on Pine River, Doyle wanted me to get him a cheap horse to haul midday lunch from the cook camp to the men in the woods, a distance of about two miles, in order to save the time of two men. I picked up a cheap one and arranged with the man I bought it from to drive it to Pembine Junction. From there I shipped the horse to North Crandon, about a two hour run from Pembine, over the Soo Line. When the time came to pay the freight charges, I found they were more than I had paid for the horse—double what they should have been. I refused to pay them or to unload the horse.

"You can wire Superintendent Pennington," I told the agent, "that I'll buy him a saddle and he can ride the old nag to hell!"

The railroad company had no use for the horse. Since they had to feed it as long as they kept it, which they didn't like, I got a rebate in short order and took it off their hands.

I had a little more trouble with the railroad company when we needed a stock chute to unload our horses at North Crandon. As we couldn't get the railroad to build one, I tried it myself. Being inexperienced in the building of railroad stock chutes, I got it a bit too close to the track and the first freight train that came in on the siding where it was built ran into it and was almost wrecked. The next day a railroad construction crew came in and built a stock chute. After that little incident I got what I wanted from the railroad company when I wanted it. I suppose they were afraid that if I didn't get what I asked for I might try to build it, with disastrous results for all concerned.

During the time we logged on Pine River, I usually visited the camps about once a week, driving in from North Crandon on the supply road. On one of these trips, in zero weather, I noticed smoke rising from a spot about a mile ahead of me on the straight corduroy road which stretched across a swamp. I couldn't imagine why there would be a fire in such a place. Curious to find out, I hurried the team along and when about eighty rods from the place smelled the strong, penetrating odor perculiar to skunks. The horses caught a whiff of it at the same time. They stood up on their hind legs and I had a devil of a time holding them to their course. When I got up to the fire I found an Indian, his squaw and papoose gathered around it. The papoose was wrapped in a bright red blanket and seated on some fine boughs on the sunny side of a great rock. It appeared to be very comfortable in spite of the zero weather. The Indian and his squaw were busy getting breakfast. They were making pancakes and I think they must have been using skunk oil for frying, which accounted for the abominable odor all around. The Indians have a great many uses for skunk oil and the odor does not seem to bother them in the least.

Another Indian family lived close to the supply road leading into the Pine River Camps. The squaw was sick and they had very little to eat—only what the Indian could kill with his gun, and that was not nearly enough for them to subsist on. Every week when I drove into the camps, I had the cook at the warehouse in North Crandon fix up a bunch of cooked provisions and I took it with me and left it behind a tree where the Indian could find it. Sometimes he would meet me at the road when he expected me and jump up and down and laugh happily and excitedly when I gave him the supplies. As a general rule, Indians will never come to one's house or camp to beg. They will accept what you give them with the joyous gratitude of children, but their inherent racial pride prevents them from asking for anything. If you treat them square, they are always your friends and they will divide anything they have with you if you are in need.

Many years ago, when I worked in the woods with a number of Menominee Indians, I often had them tell me about the feud between the Chippewas and Menominees along the Menominee River years before. The Chippewas were downstream from the Menominees and they built some dams which prevented the sturgeon from going upstream, where the Menominees could fish for them. The Menominee chief sent his son downstream with a message to the other tribe in which he requested that they open the channel and let the sturgeon up. Disregarding the courtesy due such a messenger, the Chippewas seized the youth and inflicted upon him such unique and terrible torture that it may not be described here. It was a trick worthy of Torquemada himself and when the chief's son returned home to his people, he was almost dead. That started war in earnest and for months the conflict was waged. Finally, the Menominees whipped the Chippewas, drove them for miles, and had all the sturgeon they wanted. The two tribes were bitter enemies of each other for many years, but are now reconciled.

The Farrell Dam Camp was about twelve miles from North Crandon and the "Scout's" camp was about three miles beyond. I often had to walk this distance into the camps instead of driving. When I did, I would take a good large lunch with me and stop at a halfway point which was called the "poorhouse" by the supply teamsters. There I would eat my dinner, making a little fire out of dry twigs and brewing hot tea with water from a little stream which ran by the spot. I walked into the cook shanty at the Farrell Dam Camp one evening, after hiking in from North Crandon, and found a woman sitting there crying with a six weeks old baby in her arms. The cook, William Duffran, a bashful German, was paying no attention to her. I questioned her and she told me that she had to see her husband, who was working in Camp One for Doyle; that she had been living with her aunt, but that they had quarreled and she had left. I had the cook get her a warm lunch and told the chore boy to build a fire in the office where I installed her. She had walked six miles through soft snow and she was soaked to the knees. I gave her some of my dry woolen socks and told her to take off her wet shoes and stockings and put on the dry woolens. It was really a wonder that she ever arrived at the camp. The trail by which she had come led across the north branch of Pine River and the only bridge available was a tree which had fallen across the stream. Walking across the river on that log was nothing easy for a man and it was quite a feat for a woman with a baby in her arms. When she was comfortably fixed in the office, I hiked on to Doyle's camp; when the men came in at six o'clock from the scene of the cutting two miles out, I told her husband about it. He ate his supper, walked over to the Farrell Dam Camp and the next day took his wife out to Crandon and made arrangements for her to live there. The following summer he began running around with some other woman, I heard, and his wife shot him to death in approved present-day manner. They were Kentucky mountain folk and that perhaps accounted for her handy use of the gun.

It was seldom, however, that women were seen around the logging camps. I was in bed at our headquarters at North Crandon one day when "Maude," a woman of wide but not worthy reputation in that region, blew in and told the warehouse man that she wanted to go to the Farrell Dam Camp and see her "husband," a Frenchman. I was in the next room, overheard the conversation, got up, and went out and talked to her. I knew that the man she referred to wasn't her husband, so I wouldn't let her go into the camp, but told her I'd send him out and I did later. This wild idea of hers was evidently the result of a little spree. She was still a bit boisterous.

"I got drunk as hell last night on squirrel whiskey," she told me in her shrill voice.

"Squirrel whiskey?" I echoed, for this was a new one on me, "What's that?"

"Why," she explained, "that's the kind of whiskey that makes a lumberjack run up one side of a tree and down the other side, just like a squirrel!"

So the interview was not entirely a loss. My vocabulary was richer by a choice and expressive term.

In the spring of 1895, when I dammed Spruce River and ran its waters through a canal into the lake at the head of the east branch of the Fence River, we had some trouble on the west branch which was typical of the friction so often experienced when two companies were driving on the same stream. We usually drove all the logs on the Fence, taking down those of other companies for a set rate per thousand feet, and that spring we had several million feet to drive on both branches for a number of different companies. One firm had landed about four million feet on the ice on the little lake at the head of the west branch. They wanted us to drive the logs for them, but in order to get the job done for half price they claimed they had only two million feet landed. On that basis they offered us the job of driving at $2.00 per thousand, the usual rate, or $4000 for the entire job. We scaled their logs on the ice where they landed them and knew they had four million, so we refused to drive them —as $1.00 per thousand wasn't a paying proposition. Thereupon they employed a driving crew and foreman and prepared to take down their cut themselves.

Flannigan had charge of our drive on the west branch, with "The Scout" and "The Bull" as foremen under him. I was pretty well occupied on the east branch, damming the Spruce and cutting a canal through the hogback. A man named Charles Shields was superintendent of the other company and his drive foreman was a fellow called James Black. Trouble began to brew as soon as our drive got under way. Shields and Black had charge of the reservoir dam at the head of the west branch and they did everything possible to discommode us, running the water when we didn't need it and holding it back when we did. But in spite of this attempt to defeat us, we somehow succeeded in getting our logs out of the west branch and into the main Fence. There we left them and took all our men on the east branch to finish the excavation of the canal.

When Shields and Black started their own drive, they felt the loss of the water they had wasted, and their logs were hung up for lack of it—stranded high and dry. They had to have water, so they blew the bottom out of the west branch reservoir dam and got enough to run their logs into the main Fence behind ours. They wanted to go on with their drive immediately, but I stopped that by dropping the gates of all the dams on the west branch and putting men with rifles on each one with orders to guard them until we got out of the east branch. Shields and Black couldn't move their drive an inch without the use of the dams above. They were hog-tied at an impasse, and they toed the line, for they couldn't do anything else when I had armed men on the dams guarding them. Shields and I met and within an hour we made arrangements to join drives. Each of us put on a certain number of men in the driving crew. In order to get even with him for lying about the amount of timber he had landed, I told him I'd put on about twice as many men as I did.

During the heat of the affair "The Scout" and "The Bull" raised some hell. I forget just what it was, but the opposition got out some warrants for them. The sheriff of Iron County came out from Crystal Falls to get them. "The Bull" took to the woods, but "The Scout" was arrested and taken to Crystal Falls, where he had to spend a night in jail. We got him out the next day by putting up a bond for his appearance in thirty days. When the sheriff had taken

"The Scout" away, "The Bull" came out of the woods and went back to work. No effort was made to catch him after that. After the drive was completed, we had a mock trial and the affair blew over. Mr. Frederick Carney, Sr., father of one of our partners, was a close friend of one of the owners of the other company and they settled the mixup. The other company had to pay for the dam which it had blown out.

Like all working men, riverdrivers were very particular about their eating and when grub wasn't given them on time, there was apt to be a rebellion. I had a pretty narrow escape one year when we were driving on the Fence at the hand of a bunch of men who were sore over missing a meal. The drive was laid up between two dams, water was getting scarce and the logs had to be taken down to the next dam pond on the last flood of water from the dam above. Unfortunately, the gates of the dam above were lifted very early in the morning and Doyle and Hamilton routed their men out at an early hour and put the two crews to work. They reached the rear of the logs just before the flood water and then they had to work like the devil without a letup until all the logs were floating in the next dam pond. Through some mismanagement the men didn't get their regular nine o'clock lunch and had nothing to eat until noon. When they did get their grub, twenty of the men quit and went into Witbeck to get their time checks. It was the accepted rule at that time that when men quit before the drive was completed, their wages were cut twenty-five cents a day. When the twenty found they had received this cut, they started back for Dam Three with the intention of making me raise their pay checks. I was resting in the office at Dam Three when the bunch of them arrived at the door with a one inch rope in their hands with which they intended to hang me, if I didn't raise the figures on their checks. A Winchester repeating rifle, with a full load in it, hung over the bed where I was lying. I grabbed it, leveled it at the door, cocked it, and swore by all the powers of earth, heaven, and hell that I'd shoot the first man to cross the threshold.

"There'll be one of the largest surprise parties in hell the devil ever heard of," I told them, "if any of you dare to come through that door."

They thought the matter over and decided that I meant what I said. A good hearty bluff backed by a serious looking gun was enough to make them forget their blood-thirsty intentions and they started back to Witbeck. Shortly after they left, I had the chore boy take my horse and go after them. He told them to come back—that I wanted to see them—and when they returned, I raised their pay checks and had the cook give them a good warm lunch. Then they left in the best of humor. The whole affair was due to carelessness and lack of foresight on the part of the foremen. They should have held the water in the upper dam until the river drivers had their nine o'clock lunch and then the men would have willingly worked until four or five o'clock in the afternoon without any more grub.

The last winter of Cleveland's last administration, 1895-96, [1896-97] was a mighty hard one throughout the lumber industry in northern Wisconsin and northern Michigan. Many of the lumberjacks were unable to get employment and they built shacks close to the lumber camps and stole provisions from the camps at night or during the day while the men were in the woods. Wages were very low, ranging from eighteen to twenty-six dollars per month with board and bunk. As a result of the scarcity of work, there was no kicking or quitting, as every man knew there was a dozen waiting for his job and his chances of getting another were mighty slim.

In addition to our regular activities on the Fence River that year, where we operated two large camps and logged fourteen million feet, Fred Carney, Jr., and I bought five million feet of standing timber from the Michigan Iron and Land Company of Marquette and logged it. This timber stood near Lake Michigamme and we landed it on the ice on that lake. Lake Michigamme is a body of water about six miles in length out of which flows the Michigamme River and it is very rough and treacherous in windy weather. In early days it used to take the loggers six months to get their logs out of the lake. They would drive them into it from the Peshakey River and then it was a summer's work to pick the logs up and get them over to the outlet.

We had no intention of doing our job in any such slip shod manner as that. Our five million feet were landed in one bunch on the ice and when the camp broke up in March, we hauled green timber enough for boom logs. We used the best one half inch test chain to fasten the boom logs and boomed the entire five million feet in one bunch. Some of the saloon bums threatened to cut our booms, so that the logs would go adrift and they would have a summer's work helping to pick them up and a nice, fat check to donate to the suffering saloon keepers. To prevent that I put two watchmen on the raft, one to keep guard during the day and the other during the night, both armed with Winchesters. One of the bums had an extravagantly high opinion of his fighting ability and he threatened to cut the booms and lick both of our watchmen. He tried to fulfill his threat but Tim Hogan, one of our watchmen, caught him and licked hell out of him. The bum should have known better. The name, "Tim Hogan," should have been enough to warn any sensible man. There weren't many who could have got the best of him. I picked my men with care for such posts.

When our five million feet were boomed, we got our square timber and built two large, solid rafts. On one of them we built a camp for the men and the cook and equipped it with stoves, fuel, and bedding. On the other we built a large windlass and a shelter for two horses. Further equipment consisted of two two-hundred foot lengths of two inch rope and two two-hundred pound anchors.

When the ice had released its grip on the lake and everything was in perfect readiness, we started across. There were two sweeps on the windlass and each one was worked by a horse. The two rafts, of course, were fastened to the boom of logs. One end of one of the two hundred foot lengths of rope was fastened on the windlass and the other end to one of the two hundred pound anchors. Then the anchor was taken out as far as the length of the rope allowed and dropped. Immediately the horses were set to work turning the windlass. The anchor stuck on the lake bottom and as the windlass was kept turning, the two rafts and the boom of logs were drawn slowly but irresistibly toward the anchor. While this was going on the other anchor, fastened to the other rope, was being taken out by the rowboat. As soon as the first length of rope had been pulled in, it was taken off the windlass and the other rope attached. Thus progress was almost constant. We had a favorable wind at our back, which did much to help us along, and we crossed the lake and were at the outlet within twenty-four hours, having accomplished in a day what had previously taken months to do.

Another attempt to cut our booms was made while we were crossing. Darkness fell shortly after we got under way and one of the bums sneaked up on the rear end of the boom in a small boat. It was very dark, but the raftmen were on the alert and they caught sight or sound of him and captured him. They threatened to drown him, upon which he grew very submissive. He begged off and they let him go. That was the last ever seen of him.

That winter of 1895-96 was a hard one for priests and ministers as well as lumberjacks. There was a Dutch priest named Father Lanhart who had a small parish at Michigamme village, made up mostly of poverty stricken French people. He was pretty hard up and came to me for money one day on the ice on Lake Michigamme, where our logs were landed.

"Give me ten dollars," he said, "and when you get your logs started, you'll have all the rain you want!"

"Will you give me a written guarantee to that effect?" I asked, jokingly.

"You will have the rain," he replied.

"Well," I laughed, "if I don't get the rain, there'll be merry hell to pay!" Then I gave him a ten spot and dismissed the matter from my mind. But the priest evidently didn't forget his part of the bargain. We were barely across the lake and had broken the boom when it started to rain. It rained continually, day and night, for several days.

The water in the lake rose so high that the logs were running all through the elm flats from Lake Michigamme to Republic and we were having a devil of a time getting them out of the woods. I walked over to Champion, a little mining town, and wired Carney at Witbeck, telling him to go to the Dutch priest at Michigamme and give him ten dollars to have it stop raining. I told him that I had given Father Lanhart ten dollars for rain and that we were getting too damn much of it. Father Lanhart is at Iron River, Michigan, now, and if you should need rain, ask him for it. He produces results!

In spite of the heavy rains we got all our logs safely out of the lake, ran them down the Michigamme River, and delivered them in the main Menominee, where we sold them for $10.50 per thousand, grossing over fifty thousand dollars. We had bought the timber for $25,000 and we cleaned up a good bunch of money due to the wages being exceptionally low and the supplies correspondingly cheap.

Chapter XIII Bears and Other Beasts

As I lived almost continually in the wilderness, I came into constant contact with all the birds and beasts which made it their home and these contacts often resulted in curious and amusing experiences, a few of which I shall relate here. Somehow I always seemed to have a "pull," to use modern slang, with animals of all sorts, probably because they instinctively realized and reciprocated my liking for them. This bond, if such it may be termed, between myself and members of the animal kingdom, was not peculiar to domestic animals alone. This was illustrated by one of the most unusual experiences I ever had with an animal.

On a spring drive once on the Waupee River, a tributary of the Oconto River, we had a dam go out on us and were forced to rebuild it before we could continue our drive. As quickly as possible we got our timber work in place and began running the gravel to it in wheelbarrows as rapidly as we could. We were taking the gravel from around a pine stump beneath which a mother woodchuck had her nest, occupied by a family of six. As the gravel underpinnings were taken away, the stump tumbled down, exposing the woodchuck's nest and putting her family in imminent danger of death. Immediately, the mother chuck raced over to where I was standing and threw herself at my feet, mutely but unmistakably imploring me to save her offspring from destruction.

Superficially, it was a rather amazing thing, but to one with any knowledge of animals and their ways it was not a matter of great wonder. I was the foreman of the crew,

167

in charge of the operations, and I do not doubt but what the desperate mother chuck realized this in so far as its mental processes allowed. More than that, I was the largest man in the crew, standing six feet and three inches in my stockings and weighing two hundred and ten pounds. Quite possibly my unusual size appealed to the poor woodchuck so greatly in need of an able protector. Whatever her reason for appealing to me was, her plea was successful. I was touched by it and stopped all the work until we got her and her family in its nest moved to a new location which was both safe and comfortable. She made no effort to interfere, but dumbly followed me to the new house and seemed very grateful for my timely intervention. It was an unheard of thing for the harassed foreman of a river driving crew to do, but perhaps I was a bit flattered by the chuck depending on me. And I don't think it lessened my standing among the men.

I wasn't quite so kind to a bear that I ran into on the Little Oconto River in the spring of 1872. I was helping drive saw logs out of the stream for the Oconto Lumber Company and one evening on my way back to camp after the day's work, I took a short cut through the woods instead of going along the river and passed by a logging camp which had been occupied the winter before by the crew of a log jobber. Upon breaking camp in the spring the cook had left a barrel partly filled with flour in the camp clearing. When I entered the clearing, the barrel was on its side and a creature which I took to be a black dog at first was squeezed head first into it as tight as a cork. Upon approaching closer I could tell that it was a bear by the long, coarse hair and the short tail. It didn't hear me for it was much engrossed in eating the flour and when I got close enough I gave it three or four good, hearty kicks in the rear end. It was a temptation not to be resisted. The poor bear was afraid to back out and face the unknown antagonist who was so violently disrespectful of Bruin's person, so it reared up on its hind legs with the barrel over its head and fore-quarters, staggered a bit, fell, and rolled for a short distance. When it finally emerged from the barrel, it looked more like a polar bear than a black one and there was flour flying in every direction. It beat a hasty and ungraceful retreat toward the woods. I was tempted to spear him in the hind quarters with my peavy stick while he was in the barrel, but in that case it would have been hurt and angered and might have stayed to fight it out. About the only time a bear will not run from a man in the woods is when it is injured or when it has cubs, if it is a female. A mother bear, like almost any other kind of mother, will protect its offspring to the end.

That incident calls to my mind an amusing bear story which I heard shortly after I first came to Wisconsin. This happened not long after the close of the Civil War. Three Temple brothers and an Irishman named Con Maher, all of whom had served four years in the Union Army, took up homesteads about two miles north of Oconto Falls. Each one had a quarter section and, as they were given credit for the time they had served in the army, they had to stay only one year on the land before receiving their patents from the government. Bears were plentiful in that region in those times and a large black one began to frequent their premises that spring. The Temple brothers decided to play a little joke on their black friend, so they took two gallons of syrup and a couple of quarts of whiskey, which was almost as cheap as or cheaper than syrup at that time, and mixed them together. The pail containing the mixture was then put in a place where the bear could easily get at it and the results awaited.

Bears love anything sweet, so when Mister Bruin found that pailful of syrup, he licked up every drop of it and was "stewed to the eyebrows," a crude but none-the-less expressive phrase. He was on a grand spree for about a week, tumbling about the woods in every direction. At the end of that time he was possessed of a tremendous appetite, so he staggered over to Con Maher's place and devoured everything in sight, including pigs and chickens. What he couldn't eat at the time, he carried off with him. A good sized bear can pick up a three hundred pound hog in its forepaws and walk off with it as easily as a darky walks out of a melon patch with an armful of watermelons. The Temple brother's poke had gone too far and after that raid one of them used his rifle on poor Bruin with deadly effect.

With the coming of every spring the garbage which had been accumulating in the logging camps all winter long would begin to emit an indescribable stench. The bears could scent the stuff for miles and they would come to the camps to get at it, making it fairly easy for anyone who wanted one to get a bearhide. An Englishman dropped in at one of our camps on the Escanaba River one June and asked the man in charge how he could get a bear. He was quite a sportsman, in his way, and was rather anxious to add a bearhide to his collection of trophies. Getting a bear at that time of year in that place was not a very difficult matter. Our man made a ladder to reach to the roof of the camp building and constructed a little platform there for the hunter to sit on. That done the Englishman climbed up to his point of vantage, first following the emphatic directions of our man and emptying his rifle. Once comfortably settled in his lookout, he reloaded and awaited the arrival of Mr. Bruin. The caretaker made him promise to again empty his rifle before descending, for he didn't want a dead Englishman on his hands, and he was well aware of the danger of climbing up or down a ladder with a loaded gun in one's hands. Then he cleared out and the Londoner settled down for a long wait. Not until just before nightfall did a bear make its appearance. It was a fine big fellow and it hungrily dug into the garbage heap. The hunter had a good broadside shot and hit it just back of the left shoulder. It straightened up, jumped grotesquely into the air, screamed in sudden pain, and fell dead. The two men skinned it carefully the next morning, buried the carcass, and salted the hide for shipment to New York. Our man told us later that the Englishman seemed the happiest and most grateful he had ever seen. He wanted to give the caretaker everything he had in the way of clothing.

There were many timber wolves in the region where we logged but, although their weird howls could be heard almost any night, they were seldom seen. In all the years I spent in the woods I saw but one live wolf, although I have heard the chilling music of the packs time and again. When Flannigan and I lumbered on Whitefish River and I got up very early in the morning and went into the woods to lay out roads, I could almost always hear some wolves howling in the distance. But I never saw any of them for as a general rule, they howl only when they are running away from one. They are quite cowardly and are to be feared only during winters when it is hard for them to find food. They will chase you only when you show fear and run from them.

The wolves were a constant menace to the deer. I have seen deer, half-crazed with fear, run up to my team of horses on the supply road and walk along with the horses, quite evidently seeking protection from the wolves or dogs which were pursuing them. When a deer is surrounded by a ring of howling wolves, it hasn't much chance for a continued existence, although a buck can sometimes put up a pretty good fight by ripping, spearing, and slashing with horns and hoofs, especially in the fall when his neck swells up and he becomes very powerful and dangerous. In the spring, when pursued by dogs or wolves, the deer would often attempt to seek refuge among the logs on the drive and would be battered to death trying to get out. When we broke up big jams which stretched for miles along the streams, we often found the carcasses of drowned deer in them. Occasionally, we would find one alive and haul it out on the bank of the stream, where it would lie for several days storing up strength enough to get away. A mother deer, if you happen on it in the woods with a very young fawn by its side, will pretend to be wounded and invite your pursuit in order to divert your attention from the fawn. If you capture the fawn, it will bleat or cry almost like a child for a few minutes and then will calm down and follow you anywhere, if you allow it to. But it is unsportsmanlike to take the fawn away from its mother and I never made a practice of it.

Years ago, "shining" for deer, that is, hunting them with a light, was practiced much more than it is now. When it

ome of the log jams are so tightly locked that dynamite is used to blow open the key logs which
re holding the jam.

A river dam and sluiceway where the logs were captured and directed over the dam and into the next waterway during the log drive.

The end of the drive. The logs are being herded into the sorting pens for tally count. The logs were driven down the rivers together, each log carrying the lumber company's mark. In the harbor, or at the river's mouth a sorting pens system and tally count made sure that each lumberman got his rightful logs, and that the drive gangs were credited with the correct delivered count.

A typical scene of a log jam. While the river men are trying to find and unlock the key log, the river can back up from the jam for miles hold millions of gallons of water in check. The pressure of the water on the log jam is tremendous and when the jam breaks the river men must run over the pitching logs to safety on the river bank.

The logway of a lumber mill. The logs would be floated up to this chain pickup in the mill pond. The chain with sharp points would grab the log and haul it up the incline to the sawyers working in the mill building.

A log sled chain and lock which was forged in the camp smithy's shop. These chains and lock were used in place of ropes or cables because there was no stretch to the chain and lock. Whe the huge, top-heavy load was bound down with chains, very little shifting of the logs coul take place. A shift in the logs while the sled was moving on the ice-bound ruts would mea disaster to the load, teamster and very likely to the teams of horses.

sees a light in the woods at night, a deer will stop and stare at it. The hunter can see the reflection of his light in the eyes of the deer through the surrounding darkness and is able to shoot it easily. Other animals, horses particularly, will stand and stare at a light just as deer will and as the hunter cannot always tell by the eyes what he is shooting at his prey is often something besides deer. When we were logging on the Nett River, we had to stop our drive once in the month of June and shift our crew over to the Popple River to get the logs out of that stream, as there was not enough water to continue the drive on the Nett. We had a four mule supply team at our camp on the Nett and, as the grass was fine, we turned the mules loose in the woods, where their weight increased and their condition improved rapidly. The woods were full of hunters "shining" deer with head-lights and they shot two of our mules. We could probably have located the offenders and should have had them prose-cuted, but we thought that such careless hunters might shine us as readily as deer or mules, so we let the matter drop.

Porcupines were always plentiful in the wilderness and they were prone to congregate around deserted logging camps where there were old, brine-soaked barrels in which salt pork had been kept. Like deer, they love anything with a salty flavor, and they would chew away at the barrels all night long. They are very simple and tame and they will come into one's tent and prowl around without doing a bit of harm if they are not molested. But, as everyone knows, it doesn't pay to touch them for their quills are very per-sistent about staying, once they get into live flesh. They are perfectly harmless, however, if left alone and, aside from their abnormal appetite for things salty, which sometimes is disastrous for sweat-soaked harness, they are not destructive. It is an unwritten law of the woods that a porcupine shall not be killed for mere sport. This is because it is about the only woods animal which can be killed with a club. If a man is lost in the woods and starving, it is possible for him to get a porcupine and eat the meat. Although I have never eaten it, I understand that the meat is much like chicken if parboiled several times to destroy the strong wild flavor.

And now, quite properly at the end of the chapter, we come to that little animal which has caused more headlong, ignominious flights and has inspired more fear, horror, and repulsion in the human breast than the most ugly and ferocious of jungle beasts. We come, in short, to the skunk; the polecat; the striped kitty. The introduction of a skunk into a motion picture at the present time is a sure method of gaining a laugh and its habits are well known so that I need not here dwell on its disagreeable mode of defense. One thing can be said and that is that they become offensively oderiferous only when the necessity for self-preservation arises. I have often had one come into a tent where I was sleeping, give everything in sight a careful survey and depart for further exploration elsewhere. When this happened, I was always careful to be very meek and still. It pays, unless one likes to be socially ostracized for a week following. Like almost all other wild animals, they are harmless enough if unharmed. But they do have abnormal appetites for chickens and they will raid and rob a chicken coop as neatly and cleverly as our lumberjacks cleaned out Mrs. Brown's at Turin, Michigan.

Chapter XIV Human Nature in the Woods

Not so many years have passed since the greater portion of the immensely wealthy timber and mineral lands of Wisconsin and Michigan were owned by the federal government. In the gradual change from public to private ownership, which lasted several decades, there was experienced one of the greatest periods of graft and exploitation of public resources that this nation has ever gone through. The rule was to beat the other fellow before he beat you and it was followed with a conscientiousness born of greed and avarice. The railroads and the lumber companies reaped the profits. Huge grants of land were given to the railroads for building their roads through virgin territory and opening up the new regions. This land they sold and their proceeds were often sufficient to pay for their expenditures a dozen times. The lumber companies bought a great share of their timber lands from the government at a price which made it possible for them to realize an enormous profit. They also bought from the railroads at ridiculously low prices.

When the land was open only to homesteaders, the methods of the exploiters became more devious, but just as effectual. Much of the timber and mineral land of northern Michigan and Minnesota was robbed from the government in this ingenious way. The lumber and mining companies would put lumberjacks on the land, pay them thirty dollars per month and board, keep them there for five years, and then the jacks would deed over the land to the companies. Each person was allowed to take a quarter section and the law required that certain improvements be made on the land

175

by the homesteader in order to prove up. Every homesteader had to build a house with at least two windows in
it. The crudest sort of shacks were constructed, two holes
were made in the walls for window openings and two or
three whiskey bottles stuck in each opening for glass. When
a young fellow under twenty-one wished to take a claim—
and there were many who did—the representatives of the
lumber or mining companies would take the minor to the
land office, mark "21" upon the floor with white chalk, and
have the youngster stand over the chalk mark and swear
that he was over twenty-one.

In order to get away with such practices it was necessary
to have the whole hearted coöperation of the government
land agents. Such coöperation was, as a rule, easily obtained
by sharing the fruits of graft with them. There were, of
course, some agents who were really honest and these were
the flies in the grafters' ointment. When an honest agent
made impossible the perfect consummation of the grafters'
plans, every length was gone to in order to get rid of him.
An incident happened at Ashland in the early days which
was typical. Several land sharks in Ashland, which lies at
the very top of Wisconsin, were bent upon obtaining some
valuable timber land, several million feet of pine, on an Indian reservation near there. They wanted concessions from
the government land agent, but he was a man of integrity
and they couldn't tempt him. That being impossible they
decided to get rid of him. It didn't take them long to ac-
complish their aim.

A very attractive woman of the demi-monde was im-
ported from Chicago to "vamp" the agent. She was a good-
looker and a smooth worker and it wasn't long before the

luckless land agent was badly smitten. When the affair had
reached the proper stage, she invited him up to her room.
The agent, all unaware of any plot against him, accepted
the invitation. In due time, when the siren and her victim
were in a hopelessly compromising situation, the door was
violently kicked open and there appeared in the doorway a
big, burly fellow with a gun in each hand.

"What in hell are you doing in my wife's room?" he demanded of the startled land man.

Then and then only, when it was all too late, did it dawn
on the agent that he was the victim of an elaborate plot. He
wasted no time in explanations, but left the premises in a
hurry, gathered his clothes together, and took the first train
out of Ashland, where he was seen no more. I don't imagine
he made a report of the occurrence to the Secretary of the
Interior. The next agent, no doubt, was more open to suggestion in the matter of concessions. The affair of the clever
two-gun man prospered and he later went West where he
became a millionaire.

Although they were all engaged in defrauding the government, the lumber men and the mining men seldom mixed
in each other's affairs. They were probably afraid of getting
beat if they got out of their own particular field. The Chapin
Iron Mine at Iron Mountain, Michigan, was for sale in 1892
for $100,000. It was a property generally acknowledged
to be worth many times the price asked for it, but although
there were many lumber companies which could have bought
it and knew of its value, it went a-begging for a buyer.
Finally, R. C. Flannigan of Norway, Michigan, got an op-
tion on the property, went to Cleveland, Ohio, and there sold
it to Mark Hanna for $100,000. After a short time Senator

Hanna resold it to the Federal Steel Company for $4,000,-000. The lumber companies could have bought it and made a bunch of money, but they were lumber men, not mining men.

Many were the frauds perpetrated by companies professing to be engaged in great public improvement works. They always somehow got enough from the government to pay for the improvement about ten times. One such was a certain ship canal company. It was granted great tracts of land and they sold these lands to loggers. Homesteaders had squatted on the land and when the loggers came in, the homesteaders shot their horses and did everything possible to prevent them from beginning operations. One of the homesteaders was a huge woman who was a wonderful shot with the rifle. She proved her marksmanship when she shot the clay pipe out of the mouth of one of the invading loggers.

In the days when it was possible for the lumber companies to enter government land, many poor men used to look up land with valuable timber on it. Having no money with which to enter it themselves, they would go to the lumber companies, give them a description of the land, ask them to enter it, and take a three quarters interest in it. The officials of the lumber companies would explain that it was impossible, that the timber was too far up the river, or give some such excuse. Then, when the helpless poor man had gone away disappointed, the lumber companies would enter the land at the land office by wire and send the money forward by express. That was the manner in which many poor fellows were paid for the tedious work of looking timber lands.

No one was to be trusted. It was a cutthroat game of the worst sort. Away back in 1860 Anson Eldred, for whom I worked when I first came west, had a partner and was operating a large water power mill at Stiles, Wisconsin. Mr. Eldred lived in Milwaukee and had charge of the business end of operations while the partner looked after the actual work at Stiles. The partner had a crew of men in the woods locating and estimating timber lands and was buying up government land, which could then be purchased for about $1.25 an acre, with company money. That was perfectly all right, but he was entering all the choicest lands in his wife's name, and when Eldred found that out the partnership came to an abrupt end. Some time later, when the erstwhile partner's wife died, Mr. Eldred succeeded in collecting his rightful share of the money.

Several thousand acres of the Chicago and Northwestern Railroad land was offered for sale in 1875 to the highest bidder. The competition at the sale was very keen. One rich lumberman from Oshkosh was there and another from Oconto. Much of the timber lay between the Wolf and Oconto rivers and could be logged and driven to either Oshkosh or Oconto. The man from Oshkosh wanted a certain part of the timber and the Oconto man wanted another part. The Oshkosh man called the man from Oconto to one side and told him that it was no use for them to be bidding against each other and running up the price.

"Give me the description of the land you want," he argued, "and I'll bid for you. Then, when the sale is over, I'll turn the land over to you at the purchase price."

The Oconto man agreed and the land was bought, but when he went to Oshkosh the millionaire politely told him that he'd changed his mind and was going to keep all the land for himself. There is no honor among thieves.

Politics were rotten in northern Wisconsin and upper Michigan in the early days, for the old style ballot and the caucus method of nominating candidates made it possible for a powerful and active minority to direct the course of an election. Ballot boxes were stuffed and foreigners were oft-times allowed to vote without challenge. Petty politicians and office seekers would gain votes by handing money to the saloon keepers, who would treat the saloon bums and tell them to cast their votes for the particular politician who had paid for the drinks.

Years ago a Democrat's chances of earning a living in the northern timber country were pretty poor, for all the mill owners were Republicans and insisted that the workers in the mills and in the camps vote the Republican ticket. Just after the Civil War, when the big sawmill was running at Peshtigo Harbor, there was only one Democrat employed in the entire mill, a man named William Pope. As usual, when election time came, the boss wanted everybody to vote the Republican ticket. He asked Pope to do so, but Pope flatly refused.

"I enlisted in the Union Army and fought four years to free the country from slavery," he stoutly told the boss, "and now I'll do what I can to free the people in the north from political tyranny!"

"Then you'll have to find another place to work," was the boss' ultimatum. Whereupon Pope came back with an emphatic and profane rendition of that pride-saving classic:

"You can't fire me! I've quit!"

Fights of all kinds were common, but some of them became traditional. Such a one occurred in Green Bay in 1874. An experienced lumberman by the name of Munroe owned a sawmill at Mill Center, cutting his timber off his own land, manufacturing it into lumber during the summer and hauling it into Green Bay on sleds during the winter months. This Munroe had three sons, all of them husky, lusty young men who loved nothing more than a real fight. It was their habit to go to Green Bay periodically and get into a few good fist fights, which they usually won. But on one occasion they ran into something a little tougher than usual and got the worst of the scrap. It irked them to be beaten and they wracked their brains for some method of retaliation. A few days after the scrap a young man named John Carland presented himself at the Munroe mill in search of work.

"Can you fight?" the Munroe boys asked him, after looking him over.

"I'm looking for work—nothing more," replied Carland, who was twenty-two years old and about as near perfect in body as a man could be.

The Munroe boys once more looked him over, then looked at each other and grinned.

"We'll put you to work this afternoon," they said.

After dinner one of the boys produced a set of boxing gloves, laced one pair on Carland's hands and put the other pair on his own. Then they went after each other and it wasn't long before young Munroe was down and out. Then another one tried it and then the third, each experiencing the fate of the first. By that time the first victim was able to speak.

"You're hired!" he told Carland. You're just the man we want. We usually pay thirty dollars per month with board, but we'll pay you sixty and we'll drive to Green Bay Saturday and start something!"

Early the next Saturday morning the four young battlers started for Green Bay and when they reached the city there was no time lost. The fight was on from the moment they arrived. One after another they cleaned out every place where the fighters held forth, leaving in their wake a trail of bloody noses, black eyes, lacerated mouths, and bruised bodies. They finished up everything on the East Side and then crossed the Fox River and went after the bullies around Fort Howard. When the party was over, the fiddler had to be paid, and the fines and court costs almost broke old man Munroe. But he was satisfied for his boys had won out. Later the boys went north to the iron country, took options on mineral lands, became wealthy following the discovery of iron deposits on their property, and gave their father everything he wanted for the rest of his life.

As I said before, fights were much more common in those days and it was often difficult to avoid them, whether one wanted to or not. When I lived in Oconto, I had to take the noon train to Marinette on Sunday in order to make connections with the Monday morning train to the woods on the St. Paul road. I went to the Northwestern rather early one Sunday morning to catch my train and got into a mixup with a drunken bum who was hanging around the station insulting everyone he could. There were many people passing by on their way to church and the bum was slamming them at every turn as well as making a dirty crack in my direction every once in a while. It was Sunday morning and I wanted no fight, so I went into the depot and stepped up to the ticket window to purchase tickets to Marinette for Mike Flannigan who was with me, and myself. The fool bum followed me into the depot, came up close to me, and again started to abuse me. That was a little too much for me to stand, so I hit him in the mouth and knocked him across the office. He fell backwards, his head crashed against the iron rail of a seat, he rolled to the floor and lay there, quite still. I walked outside and sent Mike back into the depot to see if the bum was dead. He wasn't, but the doctor had to put eight stitches in the cut in the back of his head and I understand he drank no more after that affair.

Every once in a while some lumberjack or bum would be touched with ambition and attempt to clean up a place single handed. Sometimes he was successful. Often he was squelched. Back in 1870, F. B. Gardner of Chicago, owned a large sawmill at Pensaukee, about five miles south of Oconto and twenty-two miles north of Green Bay. Some of his men, of course, made it a practice to go to Green Bay after pay day and stay drunk for about a week. One of these fellows got into a hotel on the west side of Fort Howard on one occasion. He was drunk and looking for trouble, so he announced that he was going to clean out the place. The hotel manager had a heavy iron poker under the box stove and reached down and took a firm grip on the handle.

"Do you ever play poker?" he asked the bum.

"You bet I do!" replied the drunk.

"Well," said the manager, "I hold a royal flush!" And he hit the bum a terrific blow over the head with the poker, laying him cold. He didn't die—skulls were too hard in those times—but the doctor who was called had to stitch up a long split in his scalp.

There were innumerable rotten dives in and around the lumbering towns and one of the worst of them was a place halfway between Pensaukee and Oconto in which the toughest of the tough hung out. The place was started by a man named William Nason, who ran it until his death. When he was dying, he called for a priest and one went out to his place to see him. The following day Billy was still alive and he called for a minister who likewise visited him. The next day Billy was still kicking, so he called for a bishop. But he couldn't get a bishop, so he died. He probably reached heaven anyway without the bishop.

After Nason's death the place was taken over by a Frenchman named Isaac, but its moral tone didn't improve any. On one occasion about twenty of the toughs who made it their hangout got hell roaring drunk and drove in to Pensaukee. There was a hotel and saloon there, run by a man named Baumgardner, and the toughs gathered in the saloon for more drinks. For protective purposes, evidently, Baumgardner had a loaded shotgun standing in one corner of the barroom. One of the gang took it up and pointed it at Baumgardner. His trigger finger slipped, the gun went off, and Baumgardner fell dead. It was very late at night, everybody was drunk, the place was in an uproar, and nobody was able to remember just who did the shooting. Baumgardner's wife was accused of it by some of the neighbors, was arrested, and later released. Of course, it was one of the gang from Isaac's dive, but no one ever found out just who. This happened over thirty years ago.

Florence, Wisconsin, in those days was one of the toughest towns in the north woods. It was a mining camp, and while all mining camps are tough, Florence was unusually so. Every other house was a saloon, a house of prostitution, or a gambling hell, and usually all three were under one roof. There was no semblance of law or order. The single street was nothing but a mudhole flanked with plank sidewalks and unsafe to venture into at night. Florence was headquarters for lumbermen. There they purchased their supplies and there, in the saloons, they picked up their crews.

The Mould place, in Florence, was one of the best known dives on the iron range. An old man named Mould owned it. He was badly crippled up with rheumatism, but he could saw a bit on the fiddle and sing a little. His daughter, Nina, was very attractive, one of the most beautiful girls on the range. The name "Nina Mould" was well known and oft heard in the mines, in the woods, everywhere. She and her parent ran the place, the father playing the violin, the daughter slinging drinks in her capacity as bar maid. The Mould place was typical of its kind. The doors were never locked and the place ran wide open all night and day, although it was fairly quiet during the day. Prostitutes thronged the place and found sure buyers of their wares in the persons of miners and lumberjacks. The white slave trade throve in Florence and more than one of the girls had a dissipated bully living off the wages of her sin, ever ready to beat her up if she didn't hand over every dollar she earned.

Once when we drove the Michigamme River, we finished the drive in the main river about four miles from Florence. We had to go in to Florence and we passed Mould's place. It was very quiet, for it was in the day time and things don't move lively in such places until night. The men all got by without being hurt, killed, or led astray. Fortunately for them they had no money and I had none to give them.

The drinking and the wild life which were so common were often productive of amusing situations. Sometime in the eighties, I forget just which year, there was a great blizzard which lasted three days and tied up the north country with three feet of snow. About four hundred traveling salesmen were stalled in Green Bay. They proceeded to get uproariously drunk and persuaded a good many Green Bay men to join them. Sunday came along when the hilarity was at its height, and all of them went to the churches. They were very welcome there in spite of their condition, for when the contribution boxes were passed around, each man put in a dollar or more, sometimes a five dollar bill. In one church each one was given a bible and the minister opened the service.

"I take my text from Genesis this morning," he announced, and all opened their bibles to the appointed place. One of the salesmen had a little better jag on than his fellows and he couldn't locate Genesis.

"Genesis?" he repeated, in a puzzled, loud, bass voice, which was heard all over the church, "Now, where in hell is Genesis?"

Everybody roared, the minister included. When things had quieted down a bit the minister smiled and said: "We will now proceed without the aid of our ungodly friend."

The lumbering regions of the lake timber states during the last half of the nineteenth century constituted about as tough and turbulent a frontier as this country has ever known. They were peopled by hardy, uncouth, and courageous men who had little respect for any law but that enforced by fist and foot. Violence of every kind was common.

Mysterious disappearances were frequent in those hectic days. There was the case of young Brooks Pendleton, brother of C. T. Pendleton for whom I worked so many years. Away back in the fifties, before the Civil War, young Pendleton left Oconto Falls one day with $1,500 on his person bound for Oconto. He was never seen again. An old man lived "in a house by the side of the road" and it was believed by many of the settlers of that vicinity that he had murdered Pendleton for the money he carried on him. His house and the surrounding country were carefully searched by people from Oconto and Oconto Falls, but no trace was found of Pendleton or his money. When the old man was on his deathbed some years later, he said that he had something to confess, but his wife wouldn't allow him to speak. It was, of course, generally believed that he was going to confess the murder of Pendleton.

Another mysterious affair was the disappearance of a man who had worked for us in the woods and on the river as foreman and was a very capable boss. After leaving us, he went into the employ of the Hiles Lumber Company and was engaged as a foreman over a road cutting crew. Immediately after lunch one noon, he shouldered his axe and went into the woods to blaze the course of the roads. At quitting time in the evening he failed to return and the next morning he was still absent. Searching parties were organized and began to hunt for him. Twenty-five timber cruisers were sent out from Oconto and they went over the whole township, but no trace was found of him. It had snowed about two inches the night following his disappearance and that made it doubly difficult to discover any trace of him.

Many people accused the Kentuckians who lived in that locality for murdering the man but, while they would have done it quickly enough if they didn't like him, it was always my belief that he committed suicide. He was suffering terribly from inflammatory rheumatism in his limbs and I believe it drove him to the point where he wanted to kill himself. There was a shaking bog within a short distance of where he had been working. He could have waded into it and sunk out of sight forever within five minutes. And he probably did.

Contrary to the popular saying, murder does not out, there were many instances to prove it in those early days. Such was the gruesome find of Hunter Orr. He was land looking and ran across the weather bleached skeleton of a man with an old, half-rotten rope still around the neck. There was no tree in the immediate vicinity and it was evident that the hapless victim of violence had not been lynched, but had been murdered and dragged to the spot.

It was about all a man's life was worth to carry any amount of money with him, unless the utmost secrecy was used. To be possessed of money was to invite thievery and sometimes violence, and it was amazing how quickly the news of possible loot traveled among questionable characters. An unfortunate incident happened at Oconto in 1870, which would have been amusing had it not been so pitiful. There was a laboring man of Norwegian extraction by the name of Thomas Thompson living there. "Norwegian Tommy," as he was popularly called, was a decent, honest fellow and a steady worker. We all liked him very much. He had a girl back in Norway and he was carefully saving up money so that he could go home and marry her. He put away every dollar he earned and even picked up cast off clothing and wore it to save money. When his savings had reached a total of about $5000, he began to worry about it and foolishly talked the matter over with a friend, who advised him to bury the money under a pine stump about a mile from town. Tommy fell for the idea and, in his innocent folly, took the friend along with him when he hid it, so that if he couldn't locate the place again the friend could. The obliging friend evidently did locate the place without any trouble, for when poor Tommy later went to look for his money, he couldn't find it and, although the friend pretended to search very diligently, the money was never recovered by Tommy. The loss almost killed him, for he had had his heart set on returning to the girl in Norway and when the money disappeared it was impossible.

It was dangerous business for an able bodied man to travel with money on him, but for a weak and aged person to do so was sheer folly. On the train between Oconto and Marinette one winter night, I butted in and broke up an attempt at theft which might have culminated in murder. There was an old man on the train who, as I found out later, was from Wausaukee and had been to Oconto with the purpose of purchasing some cedar timberland from George Byer. Mr. Byer had been out of the city and so the old man was returning home with a large sum of money still on him—about $1,500, I think—with which he intended to pay for the land. The Indians had spotted him in Oconto and were following him to Marinette in the hope that they might get an opportunity to rob him . As we were pulling into Marinette, the train came to a stop at the St. Paul road crossing, close to the roundhouse, and the two Indians grabbed the old fellow and tried to drag him out of the coach. He was no match for them, of course, and they would have experienced no difficulty in getting him out and robbing him if there had been no interference. But I was sitting in the rear of the coach near the door and as they passed me I jumped up, tore the old fellow out of their grasp and knocked one of them down and out over a seat with a luckily loaded blow between the eyes. The other Indian started for the door and I helped him through it and on his way with a swift and emphatic kick in the pants. The train had started up again and by that time was pulling into the station. We got off and, after the old fellow had told me his story, I took him to the 'Travelers' Home and told the manager to take care of him. His name was Riley; he and his sons were in the cedar lumbering business.

If a man with money on him got drunk, his chances of getting robbed were multiplied many times. The saloonkeeper's margin of profit in the old days was the money he whisked off the insensate bodies of dead-drunk lumberjacks. Such conditions continued up until the World War and, of course, are still existent to a certain extent. I was living in Marinette in 1918 and was at the railroad staion one day when a drunken lumberjack got off the train from the north and came up and spoke to me. He told me all about himself. He had just finished a winter's work in the woods near Wells, Michigan, for the I. Stephenson Lumber Company and was bound for his home in Milwaukee. He hadn't seen his family for over a year. Every pocket was filled with bills, about $600 in all, with part of which he intended to make a payment on his home. The next train for Milwaukee didn't leave until eleven o'clock that night and I knew that if he hung around Marinette that long with all that loose money on his person the saloon bums would give him some more whiskey and relieve him of the greenbacks. So I took him into a bank and made him hand over all his money to the cashier except enough with which to buy a ticket to Milwaukee. The cashier made out a draft payable to the jack and we enclosed it in an envelope and addressed it to the Milwaukee address he gave us. But for that he surely would have been robbed that night and probably would have lost his home.

I came near being held up and robbed once myself in Eagle River. We were lumbering on Popple River at the time and I had gone into Eagle River to pick up a crew of men to take down the drive. It was spring, the camps had closed down, and the little hamlet was full of drunken lumberjacks spending the winter's wages, "painting the town red." One bum felt so good he had the nerve to come up to me and tell me that he had lived off supplies stolen from our camps when we lumbered on the Fence River. I was in the railroad station after picking up a crew when a tough looking cur by the name of McBride swaggered in with a rifle and demanded some money. His eyes were red and inflamed with liquor and he was in an ugly humor. I told him I had no money to give away, which was very true. But he knew I had come after fifty men and that I probably had transportation money for them, so he cocked the rifle as a threat. I never carried a gun myself, believing that the possession of one invites violence more often than it avoids it, so I was in a pretty tight fix. Then the station agent stuck his head

through the office window and told McBride to get out of
the building. The agent could have telephoned for an officer
and the roughneck knew it, so he took the hint and made
tracks. It was a lucky thing for me that the station agent
was present. McBride hadn't figured on a third party.

During the Civil War and shortly after it was the custom to drive beef cattle from the south over the military road
which led to Ontonagon, Michigan. The cattle were driven
north during the summer and fall, grazing along the way in
the woods, and when the destination was reached, the owners
would return home on ponies and the helpers would often
hire out in the camps for the winter. The men who had
owned the cattle usually carried back large sums of money
with them which they had received in payment for the beef,
and as a result they often were robbed. One of them, returning home with a large "stake" on him, was murdered and
robbed by a "squaw man." A search was made but, as so
often happened in such cases, no trace was ever discovered
of him. When the squaw man was dying, he confessed the
murder and told how he had committed it. He had killed the
man, buried him in a shaking bog, driven his pony away,
and killed and burned his horse. I knew the "squaw man"
quite well.

A more fortunate cattle and sheep drover was the man
I have called Ferguson. He used to go into southern Wisconsin, buy up a bunch of steers, and drive them north to
fatten on his farms near Menominee. Then, when lumbermen came to buy beef for their camps, Ferguson had his
stock prepared for sale by live weight. The weight would
be largely water. He also drove up sheep from the South.
Some of these he bought, but in passing through the country

he would manage to mix with his sheep the small flocks owned by the farmers along the way, so that his flock grew like the proverbial snowball. When any of the farmers complained, he invited them to pick out their sheep, which of course they were unable to do. By taking a different route each year, he escaped being shot by the irate farmers he robbed. Ferguson was the most picturesque rogue in the north woods during my time. He stopped at nothing short of what would be sure to put him behind the bars. He throve by petty and grand larceny which would have kept an ordinary villain in jail most of his days. Yet, his personality was so engaging—both to men and women—that he evaded the law successfully to the end and died worth a quarter of a million.

In such a husky, lusty, young business as the lumber industry of the early Middlewest, there were bound to be many such crooks and rogues as Ferguson. But there were more than enough honest and able men to offset these and the names of certain lumbermen became synonymous with integrity and capability and other cardinal masculine virtues. Such men were the Stephenson brothers of Marinette and Menominee.

As a mere youth, Isaac Stephenson worked as a lumberjack for fourteen dollars per month and drove oxen for sixteen dollars per month along the Escanaba River in Delta County, Michigan. He grew up with the industry and as it assumed great proportions so did his own affairs. He had great ability and could have amassed a much greater fortune than he did—might have been worth one hundred million dollars instead of only twelve million when he died—had it not been for his great fear of getting "beat" by "outsiders." He had probably been taught some expensive lessons by dishonest men, and as a result was overly shy of all those he could not be sure of. About the only man he really trusted in a business transaction was Daniel Wells, Jr., of Milwaukee. Men as honest as the Lord, who couldn't have pulled a crooked trick had they tried, would go to him with gilt-edge timber propositions—great money-making opportunities, and he would turn them down flat.

I remember one such affair. Out in the state of Washington a company had purchased a large tract of fine timber and had built a big, first-class sawmill, but had been forced to involve itself heavily with a banking house to do so. Things didn't go as sweetly as the company had hoped and expected and its creditors soon began to tighten up. The bank had to foreclose and put all the company's holdings on the market in order to save itself. It was an exceptional opportunity for anyone who had the money to swing the deal. A Wisconsin lumberman, who was known to be as straight and honest as Washington is reputed to have been, got wind of the opportunity and went to Mr. Stephenson to try to interest him. But Stephenson wouldn't hear of it and the other fellow couldn't raise the necessary cash to pull the thing off alone. Another lumberman sold his interest in a western lumber company and bought the holdings from the bank for $250,000. Shortly after he sold the natural harbor, which was included in the holdings, for $500,000 and still had the mill and timber enough to run it for twenty-five years.

"It can't be true," insisted Stephenson when told about it and chided for his failure to take up such a promising proposition. "If it were, some of the western lumbermen would have taken it over."

Isaac Stephenson was elected to the United States Senate in 1907 as one of the two members from Wisconsin and served with credit to himself and the state for a number of years. We often came into contact with each other in a business way and in 1912, when I was starting west to investigate some lumber propositions there and look over some timberland, he gave me the following letter:

> Marinette, Wisconsin,
> September 28, 1912.

To Whom It May Concern:

Mr. J. E. Nelligan, for many years a resident of this city, is going to Mexico to look into some lumber propositions.

Mr. Nelligan has been engaged in the lumber business in this state for many years and is well known as a good and experienced lumberman. I have known him personally for years and have a high regard for him.

I take pleasure in introducing to any one interested in the lumber business a man of so much experience as Mr. Nelligan.

> Very truly,
> ISAAC STEPHENSON

Samuel M. Stephenson, brother of Isaac, was a man hard to get acquainted with and was very independent in a business way. But he was a gentleman, and we always found him honest and straightforward in our dealings with him. He was noted as an employer who was charitable to his employees and if they were ever sick or in need of fuel or provisions, he was always ready to help them out. Farming was a sort of hobby with him and in addition to his many business activities, he owned and operated several farms of thirty-five hundred acres each, close to Menominee. Dairying was his specialty and all the farms were well stocked with high grade Holstein stock. He was elected to Congress and represented his district there for two terms, identifying himself as a prominent proponent of the St. Lawrence Waterway project.

Thomas and William Stephenson, two other brothers of the family, were located at Marinette and were closely connected with the family lumber interests. Thomas supervised the operations of the two large mills at Marinette for many years while William had full charge of the main river and of all the sorting of each company's logs at the booms.

All of the four brothers were plain, unassuming men and very generous. I have known them to pick up penniless and poverty stricken children off the street and fit them out in new warm clothes from head to foot. Square-shooting, honest, and unselfish, they were typical of their time and place. It was often said of Isaac Stephenson, as it was said of almost every big employer in those times, that he was dishonest and uncharitable. I never knew him to be so. Our firm did much log driving for him and he always observed the letter of honesty in his dealings with us. As for him being uncharitable—that was nothing but balderdash. He always treated his men fairly, and when they had worked for him from lusty youth to broken down old age, he did not desert them, as did some employers, but made it a point to take good care of them to the end of their days.

Daniel Wells, Jr., of Milwaukee, was another Wisconsin lumberman who measured up to a high standard. He was acknowledged to be one of the greatest lumbermen in the state and was a large stockholder in a number of well known and powerful companies. So long as he stayed with lumber, he was all right, but he ventured into the market once and made a bad attempt to get a corner on lard—with disastrous results. It was said that he lost one million dollars in trying it. Not long after that sad affair, he visited Marinette and was shown around the properties there in which he was interested by Frederick Carney, Sr., one of his partners. Mr. Carney showed him all through the mill, tracing the production of lumber from the sawlogs in the mill pond to the graded lumber in the drying yards. The company owned a boarding house on the Island for the accommodation of mill workers, and a short distance from the boarding house a herd of fine hogs was kept, which fed off the boarding house scraps and slops and was occasionally depleted to provide pork for the table.

"Come over and see the hogs," suggested Carney, who knew of Wells' attempt to corner the lard market, when he had shown him everything else of interest.

"Damn the hogs!" replied Wells. "I never want to look a hog in the face again as long as I live!"

Colonel Balcom, one of the partners of the Holt and Balcom Lumber Company, was an enthusiastic agriculturist. The company owned a large farm on the north branch road to the Oconto River and one year they had a wonderful field of rye on it. Balcom wanted to enter some of his rye in the Cook County Exposition at Chicago, so he sent five dollars to a Frenchman who worked on the farm and asked him to pick a bunch of rye, specifying that all the stalks must be five feet or more in length.   The Frenchman couldn't find any rye in the field that long, but that didn't deter him from filling his order.   He carefully spliced the stalks and sent Colonel Balcom a very fine appearing bunch of rye five feet in length.   Balcom was much pleased and proudly presented the bunch to the judges of the exposition.   They were a little suspicious over the unusual length of the stalks, so they inspected the bunch closely and finally pulled apart the spliced stalks, much to the surprise and discomfiture of Colonel Balcom.   On returning to his farm, Balcom's first inquiry was about "the damn Frenchman who got my five dollars and spliced my rye."

It was truly amazing the number of men who worked their ways to the top of the lumber industry from poverty stricken beginnings, making their opportunities and seizing them as they went along, continually proving that great success depends only upon the will to succeed and a certain amount of innate ability.   There was J. M. Longyear of Marquette, Michigan.   He started his career as a penniless youth in the woods with a pack on his back, cruising timber and iron lands.   A few years ago, when he died, he left an estate worth close to one hundred million dollars.

Frederick Weyerhauser was born in Germany and came to this country as a lad.   One of his first jobs was working in a sawmill for the magnificent wage of one dollar per day, and from that beginning he worked his way to the peak of the industry and became one of the richest lumbermen in the entire country.   He was noted for his honesty in all his dealings, and if a man had a promising lumber proposition, he could go to Weyerhauser for backing without fear of being taken advantage of. He started more than one man in business.

Frederick Carney, Sr., of Marinette, was another self made man. As a lad of fourteen he left home with Daniel Wells, Jr., and came to Marinette, where he got his start and won to the top of the business. He was a large stock holder as well as manager of the H. Witbeck Company and was one of the most practical lumbermen in either Wisconsin or Michigan, operating in every phase of the business, from buying timberland to selling lumber. At a time when poverty was rather common among lumberjacks and mill-workers and as a result rather callously considered by many operators, Mr. Carney was noted for his generosity and charity towards the families of his employees and towards any other who stood in need of help. Everybody was his friend and nothing better than that can be said of a man. He died on a train at Menominee—quite suddenly and un-expectedly—when returning from a fishing trip.

The list of keen and capable lumbermen who made history in the Wisconsin and Michigan woods during the last part of the nineteenth century and the beginning of the twentieth is a long one and I can hope to mention but a few of them here.

There is M. J. Quinlan of Soperton, Wisconsin, manager and joint owner of the Menominee Bay Shore Lumber Company. He was one of the "old school boys" of Mus-kegon—grew up with the industry—and is still as shrewd in the purchase of a tract of timber, or the sale of a season's cut, as any man in the state.

There is John Kernan of Green Bay, another excep-tional lumberman, who is getting along in years now and

taking life easy. He worked for S. J. Murphy and Sons of
Detroit and Green Bay during the greater part of his life
and any man who worked for Simon J. Murphy, Sr., had to
know the business. Kernan had all the woods operations
of the company under his charge. He had a mania for ex-
actitude and could tell you the cost of anything connected
with his work, from an ax handle to a team of draft horses,
by consulting his always-handy notebook.

And there is Wallace McPherson, a canny Scot who
ranks among the keenest lumbermen in the country. Mc-
Pherson lives in Milwaukee now, where I make my home,
and is in the wholesale lumber business. Like every good
Scotchman, he enjoys a good Scotch highball about once a
year—never oftener—but good Scotch highballs are hard to
get nowadays. Occasionally he gets sick—probably for the
lack of one—and I call him up on the 'phone.

"What kind of a prayer shall I say for you?" I ask him,
and he shouts back, with great gusto and good humor:

"Go to hell, you old cross-back!"

We are the best of friends.

William Holmes of Menominee, Michigan, who was for
many years woods superintendent of the Kirby-Carpenter
Company, was another unusually able lumberman. The
Kirby-Carpenter firm had two large sawmills which demanded a cut of seventy-five million feet every season and
such a cut required a veritable army of lumberjacks. Sixty
miles of standard gauge railroad—main line and branches—
were constructed by Holmes and his son, and logging operations were carried on winter and summer. Conditions in
Holmes camps were so good for the time and the grub was
so excellent that they became the headquarters for the best

lumberjacks in the Middle West. I never heard of a man having a grievance against the camps run by Holmes. He was not only open handed with his men, but he fed and took care of every tramp that happened along. One of his camps was on the main line of the Chicago, Milwaukee and St. Paul Railroad and tramps, traveling the railroad, were continually dropping in asking for handouts. Homes gave his foreman and cook orders to always give every tramp a good meal before starting him on his way.

"It's taking more grub to feed the tramps than it is to feed the camp crew!" complained the cook one day.

"That makes no difference," replied Holmes. "Keep on feeding them. Provisions are pretty cheap!"

So the tramps were always safe in counting on a good meal at that camp. William Holmes, Jr., who was his father's partner and co-worker for many years, is now located at Crystal Falls and is still in the lumber business. He is much like his father—open hearted and open handed—and is very popular. Guy Holmes, another son, is in the timber estimating business, and his word on timber is as good as a gold bond. If he hadn't been so scrupulously honest, he'd have been a millionaire long ago.

Another lumberman noted for his integrity was August Spies, for whom we drove logs several seasons. Spies had a college education and was a merchant before he went into the lumber business. He was one of the most honorable men in northern Wisconsin and upper Michigan.

So there you have a few of the honest, intelligent, and able men who made logging history in Wisconsin and Michigan, men typical of the vast majority of mill owners, and loggers. They played great rôles in the drama by which the lake state pineries were leveled to earth, but they looked upon themselves as only ordinary business men and their work as ordinary business; they were all unaware of the epic scope of that which they did. It has taken another generation, which has reaped the good and the ill they sowed, to realize that. They have passed and are passing, but they will always be remembered as splendid pioneers, as men who unmercifully bent and broke the wilderness to their wishes.

Few of Them Left

JOHN EMMETT NELLIGAN, who died Thursday night, was one of the few remaining old-time lumbermen, who operated in northern Wisconsin and Michigan during the days when "pine was king." His name goes down into history with the other pioneers of the north woods, which includes such notables as Isaac Stephenson, Nelson Ludington, J. W. Wells, the Cooksons and others.

The old-time lumbermen found one blessing—plenty of fine timber. They did not have the modern facilities of transportation and manufacturing, however, and they did not enjoy the comforts of camp life that are afforded now. When they went to the woods in the fall, they stayed until the spring drive brought them into town. There were no week-end automobile trips to the movies in those days, and radio entertainment, of course, was not yet known even to science.

But despite the rigors of those early lumbering days those pioneers seemed to be hale and hearty. Many of them lived to a ripe old age and for years boasted they had never experienced sickness. The conditions developed strong people in those days. They had to be strong in order to survive.

The Escanaba Daily Press

January 9, 1937

Requiescat in pace!